PRIVATE STANDARDS IN THE UNITED STATES AND EUROPEAN UNION MARKETS FOR FRUIT AND VEGETABLES

Implications for developing countries

Trade Policy Service
Trade and Markets Division

FOOD AND AGRICULTURE ORGANIZATION OF THE UNITED NATIONS
Rome, 2007

As far as known to the author, the information provided in this report is correct at the time of writing (September 2005). However, legislation changes, standards are regularly reviewed, new research is published and activities of organizations evolve. The author may also have missed or misinterpreted essential information. In addition, much of existing rules and legislation are presented here in very short descriptions, missing out on many details. For full and up-to-date information on legislation, rules, standards and inspection/certification systems, readers should consult the appropriate official information sources.

The conclusions given in this report are considered appropriate at the time of its preparation. They may be modified in the light of further knowledge gained at subsequent stages.

The views expressed in this publication are those of the author(s) and do not necessarily reflect the views of the Food and Agriculture Organization of the United Nations. The designations employed and the presentation of material in this information product do not imply the expression of any opinion whatsoever on the part of the Food and Agriculture Organization of the United Nations concerning the legal or development status of any country, territory, city or area or of its authorities, or concerning the delimitation of its frontiers or boundaries. The mention of specific companies or products of manufacturers, whether or not these have been patented, does not imply that these have been endorsed or recommended by the Food and Agriculture Organization of the United Nations in preference to others of a similar nature that are not mentioned.

Contents

Acknowledgements

This study was funded by the FAO-Norway Trust Fund

Written by
Cora Dankers
Trade Policy Service
Trade and Markets Division, FAO

Technical editing by
Pascal Liu
Trade Policy Service
Trade and Markets Division, FAO

Formatting by
Olwen Gotts and Daniela Piergentili
Trade Policy Service
Trade and Markets Division, FAO

Layout and cover by
Eva Clare Pearce Moller
FAO

List of acronyms

(When not specified in the full name, the country/region in which the organization is active or the organization that uses the acronym is indicated between brackets.)

ACBs	Accredited Certification Bodies (IFOAM)
ACC	Agri Chain Competence Centre (Netherlands)
ACP	African, Caribbean and Pacific Group of States
AFNOR	Association Française de Normalisation (France)
AGA	Animal Production and Health Division (FAO)
AGD	Agriculture Department (FAO)
AGOA	African Growth and Opportunity Act (United States)
AIB	American Institute of Baking
ANAB	ANSI-ASQ National Accreditation Board (United States)
ANSI	American National Standards Institute
APEC	Asia-Pacific Economic Cooperation
APHIS/PPQ	Animal and Plant Health Inspection Service/Plant Protection and Quarantine (United States)
ASEAN	Association of Southeast Asian Nations
ATCWG	Agricultural Technical Cooperation Working Group (APEC)
AUSAid	Australian Government Overseas Aid Program
BCSF	Business Coalition for more Sustainable Food
BRC	British Retail Consortium
BSE	Bovine Spongiform Encephalopathy
CAC	Codex Alimentarius Commission
CADEX	Cámara de Exportadores de Santa Cruz (Bolivia)
CARICOM	Caribbean Community
CBI	Centre for the Promotion of Imports from Developing Countries
CCFAC	Codex Committee on Food Additives and Contaminants
CCPR	Codex Committee on Pesticide Residues
CEN	European Committee for Standardization
CIAA	Confederation of the Food and Drink Industries of the European Union
CID	Commercial Item Description (United States)
CIMS	Centro de Inteligencia de Mercados Sostenibles (Latin America)
CIO	Consorzio Interregionale Ortofrutticoli (Italy)
CIPMA	Centro de Investigación y Planificación del Medio Ambiente (Chile)
CLUSA	Cooperative League of the USA
CN	Combined Nomenclatura
CNCA	Certification and Accreditation Administration (China)
COAE	Center of Organic Agriculture in Egypt
COAG	Committee on Agriculture (FAO)
CoL	Cost of sustainable Living (FLO)
COLEACP	Europe/Africa-Caribbean-Pacific Liaison Committee
COMESA	Common Market for Eastern and Southern Africa
ComSec	Commonwealth Secretariat
CoP	Cost of sustainable Production (FLO)
CPMA	Canadian Produce Marketing Association
CSD	Commission on Sustainable Development (United Nations)
CTE	Committee on Trade and Environment (WTO)

CTF	Consultative Task Force on Environmental Requirements and Market Access (UNCTAD)
DAR	Deutscher Akkreditierungsrat (German Accreditation Council)
DECRG	Development Economics Research Group (World Bank)
DFID	Department for International Development (United Kingdom)
DGCCRF	Direction Générale de la Concurrence, de la Consommation et de la Répression des fraudes (Directorate of Competition, Consumption and Repression of Fraud) (France)
DGD	Decision Guidance Document (FAO/UNEP Rotterdam Convention)
DIN	Deutsches Institut für Normung
DTIS	Diagnostic Trade Integration Studies (Integrated Framework Trust Fund)
EAN	European Article Numbering
EAN	European Article Numbering
ECA	Trade HubEast and Central African Trade Hub
ECE	United Nations Economic Commission for Europe
ECL Space	Ethical Certification and Labeling Space
ECOWAS	Economic Community of West African States
EDPs	Export Development Programmes (CBI)
EEC	European Economic Community
EFSA	European Food Safety Authority
EFTA	European Free Trade Association
EFSIS	European Food Safety Inspection Service
EGE Program	Economic Globalization and the Environment Program (Pacific Institute)
EISfOM	European Information System for Organic Markets
EPA	Environmental Protection Agency (United States)
EPE	European Partners for the Environment
EPPO	European and Mediterranean Plant Protection Organization
ERS	Economic Research Service (USDA)
ESAE	Agricultural Sector in Economic Development Service (FAO)
ESCB	Basic Foodstuffs Service (FAO)
ETI	Ethical Trading Initiative
EU	European Union
EUDN	European Development Research Network
EUREP	Euro-Retailer Produce Working Group
FACB	Freedom of Association and Collective Bargaining (ILO)
FANR	Food, Agriculture and Natural Resources Department (SADC)
FAO	Food and Agriculture Organization of the United Nations
FCD	Fédération des entreprises du Commerce et de la Distribution (France)
FDA	Food and Drug Administration (United States)
FDF	Food and Drink Federation (United Kingdom)
FiBL	Forschungsinstitut für biologischen Landbau (Research Institute of Organic Agriculture) (Germany and Switzerland)
FLO	Fairtrade Labelling Organizations International
FMI	Food Marketing Institute (United States)
FNOP	FAO/Norway Programme
FPEAK	Fresh Produce Exporters Association of Kenya
FSC	Forest Stewardship Council
GAP	Good Agricultural Practice
GATT	General Agreement on Tariffs and Trade (WTO)
GFSI	Global Food Safety Initiative
GHP	Good Hygiene Practices
GMO	Genetically Modified Organism
GMP	Good Manufacturing Practice (FAO, FDA and others)

GRI	Global Reporting Initiative
GTZ	Deutsche Gesellschaft für Technische Zusammenarbeit (German Agency for Technical Cooperation)
HACCP	Hazard Analysis and Critical Control Point (System)
HDE	Hauptverband des Deutschen Einzelhandels
HHS	Health and Human Services (United States)
IAAS	International Association of Agricultural Students
IADB	Inter-American Development Bank
IAEA	International Atomic Energy Agency
IAF	International Accreditation Forum
IAMA	International Food and Agribusiness Management Association
IAPSO	International Agency for Procurement and Services
IATRC	International Agricultural Trade Research Consortium
IBCE	Instituto Boliviano de Comercio Exterior
IBS	IFOAM Basic Standards
ICC	International Chamber of Commerce
ICCO	Interchurch Organization for Development Cooperation (Netherlands)
ICFTU	International Confederation of Free Trade Unions
IDRC	International Development Research Centre
IFAD	International Fund for Agricultural Development
IFC	International Finance Corporation
IFOAM	International Federation of Organic Agriculture Movements
IFPA	International Fresh-cut Produce Association
IFS	International Food Standard
IGPN	International Green Purchasing Network
IICA	Inter-American Institute for Cooperation on Agriculture
IIED	International Institute for Environment and Development
IISD	International Institute for Sustainable Development
ILO	International Labour Organization
IMF	International Monetary Fund
INAC	International Nutrition and Agriculture Certification (Turkey)
INNI	International NGO Network on ISO
IOAS	International Organic Accreditation Service
IPPC	International Plant Protection Convention
ISEALAlliance	International Social and Environmental Accreditation and Labelling Alliance
ISO	International Organization for Standardization
ISPMs	International Standards for Phytosanitary Measures (IPPC)
ITC	International Trade Centre
IUF	International Union of Food, Agricultural, Hotel, Restaurant, Catering, Tobacco and Allied Workers Associations
JECFA	Joint FAO/WHO Expert Committee on Food Additives
JMPR	Joint FAO/WHO Meeting on Pesticide Residues
KIT	Koninklijk Instituut voor de Tropen (Royal Tropical Institute) (Netherlands)
LDCs	Least developed countries
MERCOSUR	Southern Common Market (Argentina, Brazil, Paraguay and Uruguay)
MFN	Most-Favoured Nation (WTO)
NCBA	National Cooperative Business Association (United States)
NGO	Non-governmental Organization
NOP	National Organic Program (United States)
NPIRS	National Pesticide Information Retrieval System (United States)
NPPO	National Plant Protection Organization (United States)
NTAE	Non-Traditional Agricultural Exports
OECD	Organisation for Economic Co-operation and Development

OFGF	Organic Farming and Green Food (UNESCAP)
OHSAS	Occupation Health and Safety Assessment Series
OIE	World Organisation for Animal Health
PACA	Perishable Agricultural Commodities Act (United States)
PAIA	Priority Area for Inter-disciplinary Action (FAO)
PAN	Pesticides Action Network
PIC Procedure	Prior Informed Consent Procedure
PMA	Produce Marketing Association (United States)
PMO	Produce Marketing Organization (EurepGAP)
POP	Persistent Organic Pollutant
PPMs	Process and Production Methods (WTO)
RAFI	Rural Advancement Foundation International (United States)
RIMISP	International Farming Systems Research Methodology Network (Chile)
SADC	Southern African Development Community
SADCA	Southern African Development Community Accreditation
SAI	Social Accountability International
SAN	Sustainable Agriculture Network
SARD	Sustainable Agriculture and Rural Development (United Nations)
SCS	Scientific Certification Services (United States)
SDRN	Environment and Natural Resources Service (FAO)
SIPPO	Swiss Import Promotion Programme
SNV	SNV Netherlands Development Organisation
SPS measures	Sanitary and Phytosanitary measures
SQAM	Standardization, Quality Assurance, Accreditation and Metrology (ITC)
SQF Program	Safe Quality Food (Program)
STDF	Standards and Trade Development Facility (WTO, World Bank, WHC, OIE and FAO)
STIC	Sustainable Trade and Innovation Centre
TBT	technical barriers to trade
TRADE	Trade for African Development and Enterprise (USAID)
TRIPS	Trade-Related Aspects of Intellectual Property Rights (WTO)
UEMOA	nion Économique et Monétaire Ouest Africaine
UFFVA	United Fresh Fruit and Vegetable Association (United States)
UNCED	United Nations Conference on Environment and Development
UNCTAD	United Nations Conference on Trade and Development
UN-DESA	United Nations Department of Economic and Social Affairs
UNDP	United Nations Development Programme
UNEP	United Nations Environment Programme
UNESCAP	United Nations Economic and Social Commission for Asia and the Pacific
UNIDO	United Nations Industrial Development Organization
UNSPSC	United Nations Standard Products and Services Code (UNDP)
UPC	Universal Product Code
USAID	United States Agency for International Development
USDA	United States Department of Agriculture
WATH	West African Trade Hub
WB	World Bank
WCO	World Customs Organization
WFSO	World Food Safety Organisation
WHO	World Health Organization
WSSD	World Summit on Sustainable Development
WTO	World Trade Organization
WWF	World Wide Fund for Nature

GLOSSARY OF BASIC CONCEPTS USED IN THIS REPORT

Accreditation
The evaluation and formal recognition of a certification programme by an authoritative body.

Audit
A systematic and functionally independent examination to determine whether activities and related results comply with planned objectives. For the certification programmes discussed in this report this normally means an on-site visit to verify that the production process and/or products comply with the relevant standards. Audit is also used by buyers to mean visits to their suppliers to verify that products are produced according to product specifications and procedures as stipulated in contracts.

Auditor
The person appointed to undertake the audit.

Auditing body
The body performing the auditing part of certification. Where a certification body performs its own audits, the certification body is also the auditing body.

Certification
A procedure by which a third party gives written assurance that a product, process or service is in conformity with certain standards. (ISO Guide 2, 1996).

Certification body
An organization performing certification. Sometimes referred to as the certifier or the certification agency. A certification body may oversee certification activities carried out on its behalf by other bodies. Standard owners may set requirements that certification bodies have to fulfil before they are allowed to certify against that standard.

Conformity assessment
Any activity concerned with determining directly or indirectly that relevant requirements are fulfilled. Typical examples of conformity assessment activities are sampling, testing and inspection; evaluation, verification and assurance of conformity (supplier's declaration, certification); registration, accreditation and approval as well as their combinations. (ISO Guide 2, 12.2).

Control, control body
Terms commonly used by the trade when referring to audit and auditing body.

Equivalence
When two different standards and/or conformity assessment systems achieve the same objective, lead to the same result.

Inspection, inspector and inspection body
See: audit, auditor, auditing body.

Label

Distinctive logo or statement that indicates that a product has been produced in compliance with a standard or that provides information on certain product characteristics. Labels are intended to make provision for informed decisions of purchasers.

Brand label
Logo that indicates the brand name of the product.

Private label
Term used in the trade to indicate that retailers sell products under their own name. Also called retailer branded products.

Certification label
Label to indicate that the product or the producing company has been certified against a certain standard.

Ecolabel
Label to indicate conformity to certain environmental standards. It is often used to mean lifecycle ecolabels that are granted by schemes that assess environmental impact of products on a lifecycle basis (production, consumption and waste disposal). In this publication, to avoid confusion, labels to indicate adherence to other types of environmental standards are called environmental labels.

Nutrition labels
Labels that contain information on the nutritional value of products.

Product specifications
Specifications used in contractual agreements between suppliers and buyers. Reference may be made to standards.

Regulations
Mandatory standards and rules that are set and enforced by governmental organizations. Such regulations may refer to standards set by other organizations.

Standards
Documented agreements containing technical specifications or other precise criteria to be used consistently as rules, guidelines or definitions, to ensure that materials, products, processes and services are fit for their purpose. (ISO Guide 2, 1996).

Corporate standards
For the purpose of this report: Standards set by business, usually by industry associtions. (A document specifying requirements of a single company is normally not considered to be a standard. However, in the case of multinationals with many subsidiaries and suppliers, the distinction is less clear cut.).

Environmental standards
Standards for materials, products and production processes to ensure that negative impacts on the environment are minimal or kept within certain limits.

Food safety standards

Standards for food production, processing, handling and distribution to ensure that food will not cause harm to the consumer when it is prepared and/or eaten according to its intended use.

Labour standards
Standards for working conditions to ensure workers' rights are respected

Organic standards
Standards for production and processing of organic agriculture products.

Private standards
Standards set by the private sector. For the purpose of this report this includes both corporate and non-profit NGO standards.

Social standards
Can be used to mean labour standards, but can also include standards on other social aspects of organizations and production facilities, such as the relation with neighbouring communities or minimum incomes for farmers.

Introduction

Over the past 20 years the number of standards and certification programmes for agricultural production has grown rapidly. Producers who want to export are confronted not only by a plethora of import regulations, but also within import countries by different niche markets for which specific requirements have to be fulfilled. While the adoption of voluntary standards may grant export opportunities to farmers, they can also be considered barriers to entry for those who cannot apply them either because they are too onerous or because of the lack of knowledge about their requirements. In fact, some producers and exporters increasingly regard private standards as non-tariff barriers to trade. New and more stringent standards are being developed year after year, and there is an urgent need to determine today, and in the future, the extent to which these govern world trade.

This report gives an overview of standards and certification programmes relevant for fruit and vegetable producers and exporters in developing countries with a focus on the markets of the United States and the European Union. In addition, it gives an overview of current analytical work on standards and trade, reviews major assistance programmes related to standards and provides recommendations for further research.

The concepts of standards and certification

One of the main objectives of standardization is that all companies in a given economic sector adhere to the same standards, i.e. the same procedures or product specifications. This may ease logistical procedures, facilitate trade, prevent consumer deception and improve quality. However, improvements in quality is not an automatic result of standardization. This will only be achieved when the advocated standard is a "high" standard, i.e. the requirements are an improvement in relation to common practice.

The International Organization for Standardization (ISO), defines standards as "... documented agreements containing technical specifications or other precise criteria to be used consistently as rules, guidelines or definitions to ensure that materials, products, processes and services are fit for their purpose."

From this definition it becomes clear that standards are not only used for standardization, but also as guidelines i.e. for capacity building. Agricultural standards usually do not have the purpose of standardization per se, but are developed to improve food safety, food quality or environmental and social sustainability in various farming and agrotrade systems.

A product standard is a set of criteria that products must meet. A process standard is a set of criteria for the production process. Most private standards discussed in this paper are process standards. Management system standards are a type of process standards that set criteria for management procedures, for example for documentation or for monitoring and evaluation procedures. They do not set criteria for the performance of the management system.

Setting international standards has proven to be difficult due to the variety of circumstances that exist around the world. This is especially true for agricultural practices, which have to respond to differences in climate, soils and ecosystems and are an integral part of cultural diversity. In response to this diversity, international standards are often generic standards to be used by local standard-setting organizations or certification bodies to formulate more specific standards.

Certification is a procedure by which a third party gives written assurance that a product, process or service is in conformity with certain standards (ISO Guide 2, 1996).

Certification can be seen as a form of communication along the supply chain. The certificate demonstrates to the buyer that the supplier complies with certain standards, which might be more convincing than if the supplier itself provided the assurance.

The organization performing the certification is called a certification body or certifier. The certification body may do the actual inspection or audit, or contract this out to an auditor (inspector) or auditing (inspection) body. The certification decision, i.e. the granting of the written assurance or "certificate", is based on the inspection report, possibly complemented by other information.

The system of rules, procedures and management for carrying out certification, including the standard against which a company is being certified, is called the certification programme. One certification body may execute several different certification programmes.

Certification is by definition done by a third party which does not have a direct interest in the economic relationship between the supplier and buyer. Certification is different from second-party verification, where a buyer verifies whether the supplier adheres to a requirement.

It is important to note that third-party verification does not automatically guarantee impartiality or absence of conflicts of interest. For example, the standard may have been set by any party, e.g. by the producer or by the buyer, in which case their interests are likely to be reflected in the standard. When a standard setting body certifies against its standard, a conflict of interests may also arise. The standard-setting body may want to see high implementation rates of its standard, or have a bias against certain types of producers or processors for ideological reasons, which may influence certification decisions. Finally, certification is a services industry and certification bodies compete for clients. They fear that they might lose clients if they are too strict in the interpretation of the standard.

To ensure that the certification bodies have the capacity to carry out certification programmes, they are evaluated and accredited by an authoritative institution. Certification bodies may have to be accredited by a governmental or parastatal institute, which evaluates compliance with guidelines for the operation of such bodies set by, for example, ISO, the European Union or some other entity. In addition, standard setting bodies may accredit certification bodies for the scope of their particular standard.

A certification label is a label or symbol indicating that compliance with a standard has been verified. The use of the label is usually controlled by the standard-setting body. While the certificate is a form of communication between the seller and the buyer, the label is a form of communication with the end consumer.

Standards and trade

Much has been written about the globalization and concentration that has taken place in the retailing industry for the last 20 years and is still developing. One of the consequences of retailers' increasing bargaining power is that they can impose higher requirements onto their suppliers. These requirements not only include price and product specifications, but also apply to production, processing and transport. Some technical standards, such as those for bar-coding, have been initiated by retailers to improve logistical processes. Other requirements have been included after pressure from civil society action groups. These relate mostly to the way the products are produced (e.g. no child labour). A third "driver" of standard development has been a tightening regulatory environment, such as increased levels of liability for food companies in relation to food safety aspects. Finally, competition on quality provides another incentive to adopt "high" standards.

Many retailers have their own specifications that are communicated solely to their suppliers and of which the outer world has little knowledge. For certain categories of standards, notably related to food safety, retailers and other buyers may implement standards as a group and require third party auditing and certificates.

Voluntary environmental and social standards are mostly advocated by non-governmental organizations (NGOs) and implemented by the private sector. Verification of compliance with these standards is usually done through third party auditing and certification.

This proliferation of standards and accompanying certification systems has implications for export opportunities of developing country producers of fruits and vegetables. Standards affect areas that are of concern to many governments, such as food safety, the environment, labour conditions and market opportunities.

Contents of this report

The first part of this document provides an overview of current international agreements, national regulations and private standards governing trade in fruits and vegetables. It sets the ground for the analysis of the influence of standards on market opportunities for developing country producers and exporters of fruits and vegetables. It also provides insights into the degree of interaction between international agreements, governmental regulations and private standards. The overview starts with the international framework in which standards are set, and in subsequent chapters it reviews: general import and quality requirements; phytosanitary measures; food safety standards; sustainable agriculture standards and Good Agricultural Practices (GAP); environmental standards; organic standards; and labour and social standards.

Part Two lists a body of analytical studies by various academics and other experts on standards and trade, with a focus on the implications for fruit and vegetable exports from developing countries. The overview includes: studies on the impact of governmental regulations and private standards on trade; case studies; studies of the fruit and vegetable supply chain; studies on the economics of labelling and on markets for labelled products; discussions on standard setting and the design of conformity assessment programmes; studies on the relations between private standards and (inter)governmental standards; and discussions on policy options and technical assistance. The conclusion of Part Two highlights the areas where further research is needed.

Part Three provides a non-exhaustive overview of operational initiatives by international organizations, bilateral agencies and NGO networks that address constraints and opportunities for fruit and vegetable exports from developing countries arising from private standards.

Finally, Part Four discusses the main findings of the various chapters, draws conclusions and provides suggestions for follow-up research.

Previous work on private standards and certification, Trade and Markets Division[1]

The Trade and Markets Division (EST) of FAO, has been working on issues related to environmental and social certification, and monitoring markets for certified products since 1999. EST organized an international symposium in April 2004 with the participation of 120 representatives from NGOs, the private sector and intergovernmental institutions. The participants discussed voluntary standards and certification initiatives, including how these may benefit poor farmers, plantation workers, rural communities and society as a whole.

The report[2] of the meeting concludes that compliance with environmental and social standards results in an improvement of managerial and technical capabilities of farmers. Compliance also leads to a more rational use of inputs and may reduce pollution. For example, organic agriculture has beneficial impacts on soil, water resources and biodiversity. The adoption of standards may be a means to reduce poverty in rural communities and to ensure food security, as certification enables farmers from developing countries, especially small farmers, to enter the markets of developed countries where certified products may obtain a price premium. However, there are costs involved in implementing standards and obtaining certification, and these costs are generally borne by producers instead of being evenly distributed along the supply chain.

Standards intended to increase the safety of produced food were not discussed at the meeting. Effective food safety standards will be to the benefit of consumers, although they may increase production costs. Research has found that food safety issues a priori need not worsen agricultural export potential in developing countries. However, the re-organization of the export supply chain, mostly induced by retailer consortia in developed markets, is likely to have a significant impact on markets.[3]

[1] For further information, please contact Pascal Liu, Economist (Trade), Trade Policy Service, Trade and Markets Division, FAO. Pascal.Liu@fao.org.
[2] Liu P., Andersen M., Pazderka C. 2004. Voluntary standards and certification for environmentally and socially responsible agricultural production and trade. FAO Commodities and Trade Technical paper 5.
[3] See for example Achterbosch, T. and van Tongeren, F. 2002. Food safety measures and developing countries: Literature overview; Agricultural Economics Research Institute, The Netherlands

Part 1

Overview of standards: international agreements, national regulations and private standards

Introduction

Over the past 20 years the number of standards and certification programmes for agricultural production has grown rapidly. In addition to the numerous technical regulations on food safety, plant protection and labelling developed by national governments, the private sector has increasingly established new standards for products and processes.

This chapter presents an overview of agricultural trade standards, notably those defined by international agreements, government regulations and the civil society including the business sector. It focuses on those standards which, according to researchers and industry sources, may significantly influence market opportunities for developing-country exporters of fresh fruits and vegetables. The number of standards that may affect agricultural trade, be they public or private, is large and increases virtually every day. This overview does not claim to be comprehensive but aims to briefly describe the main standards and regulations that may have an impact on international trade in fresh fruits and vegetables. In the case of regulatory standards set by governments, in particular, this report focuses on the European Union and the United States, which are by far the largest markets for the fresh produce exports of many developing countries.

Standards are firstly classified according to the most relevant subject category they tackle, and then according to the type of institutions that define and apply them. The matrix below shows how examples of major standards that could affect international trade fall in this classification system, all of which are discussed in this document. Please note that "n/a" means that no standard is considered in this report that affects international trade within the category. Note also that some standards fall within more than one subject category. In order to avoid repetition, and when this was feasible, the author has chosen to describe standards in one subject category only, therefore readers wanting to know about specific standards are advised to find them first in the complete matrix of all reviewed standards presented in the Annex 1 page 75.

Examples of various types of standards

	Governments		Civil society	
	International Agreements	National Regulations	Corporate Private	Non-profit
Frameworks	SPS	ANSI	AccountAbility	ISEAL
Product classification	HS	General Food Law	GS1	n/a
Phytosanitary	IPPC	APHIS	n/a	n/a
Food quality and safety	Codex	EPA	GFSI	n/a
Sust. Agric. and GAP	GAP	n/a	EUREPGAP	Rainforest Alliance
Environmental	ISO 14001	n/a	UK Assured Produce	n/a
Organic	Codex	NOP	ECOCERT	IFOAM
Labour and social	ILO	n/a	OHSAS	FLO

n/a: not considered in this report

1. Frameworks for standard settings and conformity assessment systems

1.1. INTERNATIONAL AGREEMENTS

The World Trade Organization

The World Trade Organization (WTO) is an international organization that defines and arbiters over trade rules that are applicable to the bulk of the world's trading nations. It was born in 1994 out of the Uruguay Round of Multilateral Trade Negotiations and its agreements concern not only trade in goods (GATT), but also in services (GATS) and intellectual property rights (TRIPS). The WTO claims to have some 30 000 pages of agreements, annexes, commitments, understandings and decisions that specify the rights and obligations of Member Nations vis-à-vis lowering trade barriers and opening of markets. WTO texts contain no standards, i.e. the WTO is not involved in establishing standards. It houses rules that guarantee that standards should not be used as disguised trade barriers.

These WTO rules govern standards set by governments or government-related institutions. Private standards are set by non-governmental entities, including civil society organizations and private enterprises and their coalitions, which may not be challenged directly before the WTO. While governmental standards are expected to be set within the framework of WTO rules, private standards (including not only food safety but also environmental and social standards) obviously have no single internationally recognized set of rules in which they should be based.

Governmental standards can be mandatory or voluntary depending on whether compliance is required for a product to enter into a particular country. Private standards can be voluntary by consensus, in the sense that every market player agrees to adopt them voluntarily. Private standards can also be de facto mandatory if they have penetrated in the market to such an extent that players wanting to participate in the market have no option other than adopting them. Due to the perceived increasing importance of private standards on world trade, some WTO member countries have proposed that these should be discussed within the realm of WTO. The concern for addressing the impact of private standards on international trade within the realm of the WTO can be better understood in light of the increasing concentration and geographical reach of the retail industry. Countries raised their concern specifically over European retailers who have created and implemented a series of sector specific farm certification standards.

In June 2005 and for the first time in the history of the Committee on Sanitary and Phytosanitary Measures (SPS Committee), the issue of private standards was raised by member countries, reflecting the perception that in some cases private standards are acting de facto mandatory. This claim concerned the EurepGAP standard, a private sector standard applied by several large-scale retailers of the European Union. The European Union explained that it could not object to private standards that were not in conflict with EU legislation. Some countries enquired about the interpretation of SPS Article 13, which reads: "Members shall take such reasonable measures as may be available to them to ensure that non-governmental entities within their territories, as well as regional bodies in which relevant entities within their territories are members, comply with the relevant provisions of this Agreement." Others asked what recourse was available to exporting countries when faced with private sector requirements that were more stringent than governmental standards. At a meeting of the SPS Committee in June 2005, the representative of Argentina stated that:

"If the private sector was going to have unnecessarily restrictive standards affecting trade and countries had no forum in which to advocate some rationalization of thesestandards, twenty years of discussions in international fora would have been wasted."

WTO - G/SPS/R/37/Rev.11

For a better understanding of this debate, this chapter first discusses briefly the WTO rules that govern standards and regulations of its Members and then continues with relevant international standards.

The General Agreement on Tariffs and Trade

Articles I and III of GATT are the basic principles of WTO and advocate non-discrimination in trade. Article I is titled General Most-Favoured-Nation Treatment, and principally means that WTO members are bound to treat the products of one country no less favourable than the like products of any other country. Article III is titled National Treatment on Internal Taxation and Regulation and principally means that once goods have entered a market, they must be treated no less favourable than like products of national origin.[4]

The term like products has been defined in the past to mean products with the same or similar physical characteristics or end uses. This has resulted in a debate on Process and Production Methods (PPMs). The WTO allows countries to adopt trade measures regulating product characteristics or their related processing and production methods, but does not allow trade restrictions on the basis of unrelated PPMs (i.e. PPMs that do not affect product characteristics).[5]

In cases where standards, certification or labelling schemes would violate Article I or III, they could still comply with GATT rules if one of the General Exceptions of Article XX applied. Nevertheless, such exceptions should not be applied arbitrarily or be unjustifiably discriminative between countries where the same conditions prevail.

The most relevant exceptions listed in GATT Article XX are:

...nothing in this Agreement shall be construed to prevent adoption or enforcement by any Member of measures:

(a) necessary to protect public morals;

(b) necessary to protect human, animal or plant life or health;

(d) [...] relating to the protection of patents, trade marks and copyrights, and the prevention of deceptive practices;

(g) relating to the conservation of exhaustible natural resources if such measures are made effective in conjunction with restrictions on domestic production or consumption.

Technical Barriers to Trade Agreement[6]

The Agreement on Technical Barriers to Trade (TBT) deals with standards and certification in general. Under the TBT, a standard is called a technical regulation if compliance is mandatory and a standard if it is not mandatory. According to TBT Articles, technical regulations shall not be more trade restrictive than necessary to fulfil a legitimate objective. Such legitimate objectives are inter alia: national security requirements; the prevention of deceptive practices; and protection of human health or safety, animal or plant life or health, or the environment. A Member shall upon the request of another Member explain the justification of that technical regulation. The TBT encourages the

---

[4] Adapted from: FAO. 2000. Multilateral Trade Negotiations on Agriculture: A resource manual: pp. 103-104.

[5] After: FAO. 2001. Product certification and eco-labelling for fisheries sustainability. By C.R. Wessells, K. Cochrane, C. Deere and P. Wallis. FAO Fisheries Technical Paper, No.42: pp. 60-63.

[6] WTO. 1994. The results of the Uruguay Round of Multilateral Trade Negotiations: the legal texts. Annex 1a Multilateral Agreements on Trade in Goods. Geneva.

international harmonization of technical regulations and standards.

The Code of Good Practice for standards (TBT Annex 3, acceptance of the Code is optional) requests standardizing bodies to apply the Most Favoured Nation (MFN) principle and the National Treatment (NT) Principle of the GATT in their work. Furthermore, the standardizing body shall ensure that standards are not prepared, adopted or applied with a view to, or with the effect of, creating unnecessary obstacles to international trade.

TBT Articles 5 to 9 on Conformity Assessment apply to both regulations and standards. They require that conformity assessment procedures should be non-discriminatory for like products and shall not be more strict than necessary. Information requirements should be limited to what is necessary to assess conformity and determine fees. Confidentiality should be respected. In addition, a complaints and corrective action procedure should exist in each country, and the international harmonization of conformity assessment procedures is sought. Members shall accept conformity assessment procedures in other countries provided that they are satisfied with the level of equivalence.

Each member shall have an enquiry point, which has to be able to provide information on any regulations, standards and conformity assessment procedures in its territory. Members shall take account of the special situation of developing countries to ensure that no unnecessary obstacles are created to exports from developing countries.

SPS Agreement[7],[8]

The Agreement on the Application of Sanitary and Phytosanitary Measures (or SPS agreement) allows countries to set their own standards. But it also indicates that regulations must be based on science. Measures should be applied only to the extent necessary to protect human, animal or plant life or health. In addition, they should not arbitrarily or unjustifiably discriminate between countries where identical or similar conditions prevail.

Member countries are encouraged to use international standards where they exist. Annex A of the SPS Agreement specifically defines international standards, guidelines and recommendations to be:
- for food safety, those established by the Codex Alimentarius Commission
- for plant health, those developed under the auspices of the FAO's Secretariat of the International Plant Protection Convention.
- for animal health and zoonoses, those of the International Office of Epizootics
- for matters not covered by the above organizations, appropriate standards, guidelines and recommendations promulgated by other relevant international organizations open for membership to all Members, as identified by the Committee.[9]

Members may use measures which result in higher levels of protection than those defined under the above, provided the scientific justification is valid. They can also set higher standards based on appropriate assessment of risks so long as the approach is consistent and not arbitrary. In cases where relevant scientific evidence is insufficient, a Member may provisionally adopt SPS measures on the basis of available pertinent information. In such circumstances, members shall seek to obtain the additional information necessary for a more objective assessment of risk and review the measure accordingly within a reasonable period of time.

---

[7] WTO. 1994. The results of the Uruguay Round of Multilateral Trade Negotiations: the legal texts. Annex 1a Multilateral Agreements on Trade in Goods. Geneva.

[8] WTO. 2003. Understanding the WTO. 3rd edition.

[9] "Members" are WTO members. The requirement that international organizations should be open for membership by all WTO members may mean that SPS standards set by trade associations and/or international NGOs would not fall under the definition of international standards of the SPS Agreement.

The agreement still allows countries to use different standards and different methods of inspecting products. If an exporting country can demonstrate that the measures it applies to its exports achieve the same level of health protection as in the importing country, then the importing country is expected to accept the exporting country's standards and methods.

The agreement includes provisions on control, inspection and approval procedures that may be used to ensure compliance with their sanitary and phytosanitary measures. In addition, the agreement stipulates that governments must provide advance notice of new or changed sanitary and phytosanitary regulations, and establish a national enquiry point to provide information. The agreement complements that on Technical Barriers to Trade and they are equal in the WTO hierarchy.

Other WTO Agreements

The Agreement on Agriculture. The Uruguay Round produced the first multilateral agreement dedicated to the agricultural sector. It was implemented over a six year period (and is still being implemented by developing countries under their 10-year period), that began in 1995. The new rule for market access in agricultural products is "tariffs only" and many quotas have been replaced by tariffs. Furthermore, commitments have been made to reduce tariffs and trade-distorting subsidies. The Uruguay Round agreement included a commitment to continue the reform through new negotiations and these were launched in 2000. The result is that there are fewer trade barriers as result of quotas and tariffs. Thus, technical barriers to trade have become relatively more important.

The Agreement on Import Licensing Procedures defines import licensing as administrative procedures (…) requiring the submission of an application or other documentation (other than that required for customs purposes) to the relevant administrative body as a prior condition for importation into the customs territory of the importing Member. The Agreement says import licensing should be simple, transparent and predictable. The agreement tries to minimize the importers' burden in applying for licenses, so that the administrative work does not in itself restrict or distort imports. The agreement also sets criteria for automatic licensing so that the procedures used do not restrict trade.

Rules of origin are the criteria to define where a product was made, essential for implementing quotas, preferential tariffs, anti-dumping actions etc. and to compile trade statistics and for made in labels. This is complicated because products may contain ingredients from - or may have undergone processing in - several countries. The Rules of Origin Agreement requires that rules are transparent; that they do not have restricting, distorting or disruptive effects on trade; that they are administered in a consistent, uniform, impartial and reasonable manner; and that they are based on a positive standard (the standard should state what does confer origin rather than what does not). The agreement establishes a harmonization work programme that aims for a single set of rules of origin to be applied under non-preferential trading conditions by all WTO members.[10] The work was due to end in July 1998, but in July 2002, the Committee on Rules of Origin forwarded 93 core policy issues to the General Council.[11] In August 2004 the General Council extended the deadline for completion of the policy issues to July 2005.[12]

A place name is sometimes used to identify a product. Well known examples include Champagne and Roquefort cheese. The use of such geographical indication does not fall under the Rules of Origin agreement but under the Agreement on Trade-Related Aspects of Intellectual Property Rights (TRIPS). Unlike the other agreements mentioned, the TRIPS Agreement does not specify the general principles in GATT, but is an agreement on its

---

[10] WTO. 2003. Understanding the WTO. 3rd edition.
[11] WTO G/RO/52; WT/GC/M/75.
[12] WTO G/RO/59.

own, on the same level as GATT. On geographical indications, the TRIPS Agreement says countries have to prevent the misuse of place names to prevent misleading the public. Some exceptions are allowed, for example if the name is already protected as a trademark or if it has become a generic term. The TRIPS agreement also defines what types of signs must be eligible for protection as trademarks, and what the minimum rights conferred on their owners must be.

Bendell and Font argue that standardization, certification and accreditation are a services industry and may thus fall within the context of the General Agreement on Trade in Services (GATS).[13] Like TRIPS, GATS is an agreement on its own. The agreement covers all internationally-traded services. It also defines four ways (or "modes") of trading services: cross border supply (services supplied from one country to another); consumption abroad (consumers or firms making use of a service in another country); commercial presence (a company setting up subsidiaries or branches in another country); presence of natural persons (individuals travelling from their own country to supply services in another). Analytical work of what GATS would mean for standardization, certification and accreditation services is not known to the author of this report and conducting such an analysis falls outside the scope of this study.

The Trade and Environment Committee[14]

At the end of the Uruguay Round in 1994, trade ministers from participating countries decided to begin a comprehensive work programme on trade and environment in the WTO. They created the Trade and Environment Committee. This has brought environmental and sustainable development issues into the mainstream of WTO work.

The CTE states that the most effective way to deal with international environmental problems is through international environmental agreements. In other words, using the provisions of an international environmental agreement is better than one country trying on its own to change other countries' environmental policies. Labelling environmentally-friendly products is an important environmental policy instrument. For the WTO, the key point is that labelling requirements and practices should not discriminate — either between trading partners (most favoured nation treatment should apply), or between domestically-produced goods or services and imports (national treatment). Discussion in the CTE continues on how to handle (under the rules of the TBT Agreement) labelling used to describe whether processing methods are environmentally-friendly.

The Doha Round of Multilateral Trade Negotiations

At the Fourth Ministerial Conference in Doha, Qatar, in November 2001 WTO Member Governments agreed to launch new negotiations. They also agreed to work on other issues, in particular the implementation of the present agreements.

- Among the many issues under discussion, some are relevant for standards and certification: Implementation issues of the Agreement on Sanitary and Phytosanitary (SPS) measures.
- Implementation issues of the Agreement on Technical Barriers to Trade.
- Completing the harmonization of Rules of Origin and dealing with interim arrangements.
- Review of GATT Articles 5 (Freedom of Transit), 8 (Fees and Formalities Connected with Importation and Exportation) and 10 (Publication and Administration of Trade Regulations) and identification of the trade facilitation needs and priorities of Members, in particular developing and least-developed countries (by the WTO Goods Council).

---

[13] Bendell, J. and Font, X. 2004. Which tourism rules? Green standards and GATS. In: Annals of Tourism Research Vol. 31. No. 1 pp. 139-156. (Page 152).

[14] WTO. 2003. Understanding the WTO. 3rd edition

Furthermore Ministers set priorities for the discussions in the Committee on Trade and Environment. Among them were:

- The effect of environmental measures on market access, especially for developing countries.
- Environmental labelling requirements: impact of ecolabelling on trade and whether existing WTO rules stand in the way of ecolabelling policies. Parallel discussions are to take place in the Technical Barriers to Trade (TBT) Committee.

Codex Alimentarius[15]

The Codex Alimentarius Commission (CAC) was created in 1963 by FAO and WHO to develop food standards and related texts such as guidelines, codes of practice, etc. under the Joint FAO/WHO Food Standards Programme. The main purposes of this Programme are protecting the health of consumers and ensuring fair practices in the food trade, and promoting coordination of all food standards work undertaken by international governmental and non-governmental organizations.

As of January 2005, the Commission had 174 Members (173 member countries and one member organization, i.e. the European Community). The Commission meets annually. Subsidiary bodies are allocated the responsibility to develop proposed draft and draft standards which are circulated to Member Governments and observers for comments. After the Commission adopts the draft standard it is included in the Codex Alimentarius, which is a collection of internationally adopted food standards and related texts presented in a uniform manner.

Participation in all standard setting activities of Codex requires expertise and resources. To address this, FAO/WHO launched a Trust Fund for Enhanced Participation in Codex in 2003. The fund is seeking US$40 million over a 12-year period to help developing countries and countries in transition to increase their participation in the Codex Alimentarius Commission.[16]

Codex standards, guidelines and recommendations are voluntary in nature. However, since the creation of the WTO on 1 January 1995 and, in particular, under the WTO SPS and TBT Agreements that encourage member countries to harmonize their national regulations with international standards, an increasing number of countries are using Codex texts as a basis for their national food regulations. In other words, Codex standards and related texts have become international benchmarks under the WTO framework.

Codex standards fall into two groups, the general (horizontal) standards applying across commodities and the commodity (vertical) standards applying to single foods or a group of foods. Codex standards and related texts can be found on the Codex official standards search web site: http://www.codexalimentarius.net/search/advanced.do?lang=en

While a number of Codex standards apply to specific commodities, other Codex texts such as codes of practices, guidelines, etc. are directed to assist countries in improving national food control systems and complying with the provisions established in Codex standards. For example some texts give guidance on good hygienic/production practices throughout the whole processing and handling system.

Codex texts that set a framework for national certification systems include among others[17]:

CAC/GL 20 Principles for food import and export inspection and certification.

---

[15] FAO/WHO. 2005. Understanding the Codex Alimentarius. Rome.

[16] http://www.who.int/foodsafety/codex/trustfund/en_

[17] A complete list of food import and export inspection and certification texts adopted by the Codex Alimentarius Commission is available at http://www.codexalimentarius.net/ by directing to Official Standards, Special Publications.

These guidelines include definitions of terms such as audit, certification etc. They state that inspection systems for food safety should be based on a risk assessment. In general the guidelines are very similar to TBT requirements on conformity assessment, i.e. non-discrimination, not more trade restrictive than necessary, international harmonization and transparency.

http://www.codexalimentarius.net/download/standards/37/CXG_020e.pdf

CAC/GL 26 Guidelines for the design, operation, assessment and accreditation of food import and export inspection and certification systems.

This guideline also stresses the importance of the use of risk analysis. It promotes the use of HACCP. It states that governments should take into account the voluntary utilization of quality assurance procedures by food businesses and that these may influence the official control methods and procedures. It gives guidelines on the recognition of equivalence of inspection and certification systems between states. Furthermore it provides advice on the infrastructure necessary for an effective inspection system: the legislative framework; the elements of control programmes; facilities, equipment, transportation and communications; laboratories; personnel; and decision criteria and actions in case of non-conformity. Finally it gives guidelines on certification systems, the accreditation of inspection and certification bodies to provide services on behalf of official agencies, and advises to regularly evaluate the whole national system. Operations of the inspection and certification system should be as transparent as possible, both for consumers and for exporting countries to be able to demonstrate equivalency.

http://www.codexalimentarius.net/download/standards/354/CXG_026e.pdf

CAC/GL 34 Guidelines for the development of equivalence agreements regarding food import and export inspection and certification systems.

Practical guidance for governments desiring to enter into bilateral or multilateral equivalence agreements. http://www.codexalimentarius.net/download/standards/362/CXG_034e.pdf

CAC/GL 47 Guidelines for food import control systems.

Additional guidelines to CAC/GL 20 and 26.

http://www.codexalimentarius.net/download/standards/10075/CXG_047e.pdf

ISO and the IAF[18]

The International Organization for Standardization (ISO) consists of 150 national standards institutes that are either governmental or parastatal bodies or are non-governmental bodies dominated by industry representatives. ISO allows only one member institute per country. The ISO system has long been considered the major standard setting body for international harmonized industry standards and is explicitly recognized by the TBT as providing internationally accepted standards.

Many developing countries have so far not been able to participate in ISO because of the lack of a national standards institute or the lack of funds for membership. However, membership of developing countries has increased fast in the last decade. ISO has launched a Five-year Plan for Developing Countries, which aims to increase the participation of its developing country members and strengthen their standardization infrastructures.

ISO standards and guidelines for standard-setting procedures and conformity assessment procedures are being widely adopted by other standard-setting bodies and by accreditation and certification bodies. These standards have recently been updated and expanded. The most important ISO standards in this respect are standards for:
- the definitions (Guide 2),

---

[18] Although ISO is not an intergovernmental organization, its standards are widely recognized by governments and therefore they will be discussed together with intergovernmental standards across this document.

- standard-setting (Guides 7 and 59),
- suppliers' declaration of conformity (Guides 17050-1/2),
- third-party certification system (Guide 28),
- conformity assessment (Guide 53, 60 (code of good practice), 67, 17000, 17002, 17003) auditing (Guide 10011),
- accreditation (Guide 17011, formerly guide 61),
- the operation of inspection and certification bodies (Guides 62, 65, 17020, 17024, 17040),
- the recognition and acceptance of conformity assessment results (Guide 68),
- third party marks of conformity (labelling) (Guide 17030).

ISO has also developed standards for quality systems, environmental management systems and recently for food safety management systems. It is in the process of developing its first standards on social responsibility. These will be dealt with under the respective following chapters.

ISO itself does not certify companies nor does it accredit certification bodies. For this reason, the ISO logo can not be used in connection with certificates or on product labels. Certification against ISO standards is carried out by either governmental or private certification bodies on their own responsibility. Usually these bodies are required by national law to be accredited by the national accreditation authorities.

The International Accreditation Forum (IAF) is the world association of Conformity Assessment Accreditation Bodies. Similarly to ISO it admits only one member institute per country. It aims to harmonize and improve accreditation systems and members must declare their intention to join the IAF Mutual Recognition Agreement. Many companies that require certification do require certification from a body that is accredited by an IAF member against ISO standards 62 or 65.

1.2. NATIONAL STANDARD SETTING SYSTEMS

United States

In the United States various federal and state departments are responsible for technical regulations related to fruits and vegetables, such as the United States Department of Agriculture (USDA), the Food and Drug Administration (FDA) and the Environmental Protection Agency (EPA). If such regulations include certification requirements, certifications are carried out by governmental bodies or by private certification bodies that are accredited by a governmental body.

For voluntary industry standards and related certification and accreditation, the American National Standards Institute (ANSI) plays a central role. ANSI is the US member of ISO. ANSI does not itself develop standards but accredits standard developers. For example, it accredited the Agricultural Research Service of the US Department of Agriculture for developing common names for pest control chemicals.

ANSI also accredits certification bodies, specifically for product and personnel certification programmes, and it accredited one body to conduct EurepGAP certification. The ANSI-ASQ National Accreditation Board (ANAB) is the US accreditation body for management systems and IAF members. ANAB accredits certification bodies for ISO 9001 quality management systems and ISO 14001 environmental management systems, as well as a number of industry-specific requirements.

European Union

Technical regulations affecting fruit and vegetable exporters in developing countries are increasingly being harmonized across the European Union. EU Directives require conformity assessment procedures for regulations (mandatory standards) to be carried out by testing laboratories, and certification or inspection bodies notified by the Member

Governments to the European Commission.

CEN, the European Committee for Standardization, was founded in 1961 by the national standards bodies in the EEC and EFTA countries. Formally adopted standards must be implemented at national level and conflicting standards withdrawn. On behalf of governments, the European Commission or EFTA Secretariat may request the European Standards Organizations to develop standards by issuing formal mandates.

CEN may delegate preparatory standard development to its members. Some of the biggest are the British Standards Institute in the United Kingdom, the Association Française de Normalisation (AFNOR) in France and the Deutsches Institut für Normung (DIN) in Germany. All three are part of larger groups with commercial subsidiaries that offer certification or consultation services. The standard setting work is partly financed by these commercial activities and/or the selling of the official standards documents. Many ISO norms and CEN norms are the same or equivalent (e.g. ISO 62 = EN 45012, ISO 65 = EN 45011).

Due to historic differences, there still may exist differences between EU member countries in their standard setting, certification and accreditation systems. For example, in the United Kingdom and France there is just one recognized accreditation body (UKAS and COFRAC, respectively), whereas in Germany there are several accreditation bodies for different industries or regions. For international representation, in the IAF for example, the German Accreditation Council (DAR) has been formed.

1.3. CORPORATE STANDARDS

The private sector, at least the bigger companies, has a strong influence on the ISO standards through the national standards organizations. Furthermore, they form an important constituency of AccountAbility and the Global Reporting Initiative (GRI).

AccountAbility and the Global Reporting Initiative

AccountAbility is a non-profit organization established in 1996. Currently it has over 300 members in 20 countries. Members include businesses, NGOs, service providers and researchers. AccountAbility has developed the AA1000 series of standards to establish a systematic stakeholder engagement process for generating indicators, targets, and reporting systems. It does not prescribe what should be reported on but rather the how of reporting, and thus complements the GRI Reporting Guidelines (see below). For more information on AccountAbility and the AA1000 series, please see http://www.accountability.org.uk/default.asp.

Furthermore, the AA1000 Assurance standard is a standard for assurance practitioners (i.e. auditors and certification bodies, etc.). Since 2005 these practitioners can seek certification against this standard through the Certified Sustainability Assurance Practitioner programme.

GRI is a multistakeholder process incorporating representatives from business, accountancy, investment and NGOs from around the world. Created in 1997, GRI became independent in 2002 and is an official collaborating centre of the United Nations Environment Programme (UNEP) and the UN Global Compact.

GRI has developed Sustainability Reporting Guidelines. These guidelines are for voluntary use by organizations for reporting on the economic, environmental and social dimensions of their activities, products, and services. In addition to the guidelines, GRI is developing technical protocols on indicator measurement. Each protocol addresses a specific indicator (e.g. energy use, child labour) by providing detailed definitions, procedures, formulae and references to ensure consistency across reports. GRI is also developing sector supplements to address specific issues of certain industries. Currently there is no supplement for the agricultural sector.

Organizations that have used the GRI guidelines in their sustainability reporting can request to be included in GRI's database. They can also indicate if their report is in accordance with the guidelines. GRI does not verify if these claims are true.

Only two companies from the agricultural sector are included in the database, neither of which is from the fruit and vegetable sector. For more information, see

http://www.globalreporting.org/index.asp.

AccountAbility and GRI have not themselves developed certification systems for their stakeholder engagement and reporting standards. However, certification bodies have developed such systems. Certification companies such as SGS and BVQI offer certification programmes to verify if a company's stakeholder engagement system and reporting comply with the AA1000 and Sustainability Reporting Guidelines.

1.4. CIVIL SOCIETY AND NGOS

Some civil society organizations and NGOs are represented in the AccountAbility and Global Reporting Initiative discussed above. A purely NGO organization is the ISEAL Alliance.

International Social and Environmental Accreditation and Labelling (ISEAL) Alliance

ISEAL Alliance brings together leading international standard-setting, accreditation and labelling organizations that are concerned with social and environmental criteria in product and renewable resource management certification. The main goals of the ISEAL Alliance are to attain credibility and recognition for the participating organizations, to defend common interests and to promote continuing professional improvement of member activities.

Standards and accreditation programmes of ISEAL members are international in nature and focus on production and trade methods and processes. These characteristics combine to make ISEAL member organizations unusual within the fields of conformity assessment and voluntary labelling instruments. As such, members have prioritized the need to monitor and provide input into policy development. In this way, members can ensure that interpretations of regulatory issues and voluntary frameworks are favourable to member programmes, and that these types of conformity assessment programmes are recognized as legitimate.

ISEAL developed a Code of Good Practice for Setting Social and Environmental Standards through a multistakeholder discussion process, to complement Annex 3 of the TBT Agreement and relevant elements of ISO Guide 59. The code was launched in April 2004 and a first public review conducted in 2005.

A programme of peer review is being implemented for both standard-setting and accreditation. ISEAL members which run accreditation programmes are being assessed against ISO Guide 17011 (before Guide 61) on accreditation.

A long-term objective for ISEAL is to act as a broker in the harmonization of members' systems. This begins with the harmonization of procedures for setting standards and carrying out accreditation, and will move to the elimination of duplication of standards.

For more information and the Code of Good Practice, please see: http://isealalliance.org.

2. General quality and import requirements, product classification, traceability and labels of origin

There are many product classification systems and commercial product identification code systems. Classification systems are hierarchical systems of product categories used for the application of customs duties and taxes. Quality classification or grading is the process of assigning quality indications to a batch, lot or other unit of one product category.

Identification codes are neutral numbers used for traceability purposes. They can convey information on the location of production, the producing or trading company, shipping container codes, etc. In vertical and closely integrated supply chains, identification codes are of less importance (e.g. in banana production, where the growing and packaging are usually done on the same plantation).

2.1. INTERNATIONAL AGREEMENTS

The World Customs Organization (WCO) and the Harmonized System

WCO is an independent intergovernmental body whose mission is to enhance the effectiveness and efficiency of customs administrations. The 166 WCO members are responsible for processing more than 98 percent of all international trade.

The WCO developed and introduced the Harmonized Commodity Description and Coding System (or, in short, Harmonized System - HS). HS numbers are classification numbers assigned to individual products. They are typically 6 to 10 digits long. The HS number is used by customs authorities around the world to identify products for the application of duties and taxes. HS numbers are used to determine tariff rates and may be needed for various documents along the supply chain.

In June 1999, a revised International Convention on the Simplification and Harmonization of Customs Procedures (the Kyoto Convention) was approved by the Council. The revised Kyoto Convention must have at least 40 Contracting Parties before it can enter into force. In April 2004 the European Union and 12 of its Member States acceded to the convention, which brought the number of contracting parties to 31. The United States has not yet ratified the convention.

The WCO also has an official role regarding the implementation of the WTO Agreement on Rules of Origin and administers the WTO Valuation Agreement. A Cooperation Agreement with the International Chamber of Commerce (ICC) seeks to further standardize and improve the level of customs capabilities worldwide.[19]

---

[19] WCO. WCO brochure 2002 available at: http://www.wcoomd.org/ie/index.html

United Nations Standard Products and Services Code (UNSPSC)[20]

The Harmonized System cannot be used for procurement because it does not include services. For procurement and electronic commerce, the UNSPSC system is more suitable. The UNSPSC was jointly developed by the United Nations Development Programme (UNDP) and Dun & Bradstreet Corporation in 1998. It is a global classification system similar to the HS. At its core is an eight-digit classification code.

Codex Alimentarius
CAC/GL 1 General guidelines on claims
These guidelines contain general provisions on prohibited/potentially misleading claims and on conditions for use of claims vis-à-vis origin, nutritional properties, nature, production, processing, composition or any other quality.
    http://www.codexalimentarius.net/download/standards/33/CXG_001e.pdf

CAC/RCP 44 Recommended international code of practice for packaging and transport of fresh fruit and vegetables
This code recommends proper packaging and transport of fresh fruit and vegetables in order to maintain produce quality during transportation and marketing.
    http://www.codexalimentarius.net/download/standards/322/CXP_044e.pdf

Codex STAN 1: General standard for the labelling of pre-packaged food
This standard addresses mandatory and voluntary labelling requirements applicable to pre-packaged foods. The standard requires that labelling of pre-packaged food is not false, misleading or deceptive. Labels of pre-packaged foods shall contain: the name of the food; list of ingredients and declaration of allergens; net contents and drained weight; name and address of the manufacturer, packer, distributor, exporter, importer or vendor; country of origin if its omission would mislead or deceive the consumer; lot identification; and date marking and storage instructions.
    http://www.codexalimentarius.net/download/standards/32/CXS_001e.pdf

Traceability/product tracing
According to Codex, traceability/product tracing is defined as: the ability to follow the movement of a food through specified stage(s) of production, processing and distribution. Codex is currently elaborating the draft Principles for the Application of Traceability/Product Tracing in the Context of Food Import and Export Inspection and Certification Systems (these principles may be adopted in July 2006).

Codex commodity standards: many standards for specific fruits and vegetables and for groups of fruits and vegetables
There are separate standards for every form in which the fruits and vegetables are marketed: fresh, quick frozen, canned, dried, and/or as juice (single strength, concentrated etc.), including nectars and purées. There are also standards for multi-ingredient fruit and vegetable products such as mango chutney, pickled cucumbers, bouillons and consommés. On fresh products, there are many more standards for fruits than for vegetables.

There are 26 standards for fresh fruits and vegetables.[21] They mainly address quality provisions covering classification, sizing, presentation and labelling. They usually contain maximum tolerances for quality and sizing requirements, e.g. levels of bruising, rot and

---

[20] http://www.unspsc.org/Defaults.asp

[21] They are: STAN 182 Pineapple, 183 Papaya, 184 Mango, 185 Nopal, 186 Prickly Pear, 187 Carambola, 188 Baby Corn, 196 Litchi, 197 Avocado, 204 Mangosteens, 205 Bananas, 213 Limes, 214 Pummelos (Citrus grandis), 215 Guavas, 216 Chayotes, 217 Mexican Limes, 218 Ginger, 219 Grapefruits (Citrus paradisi), 220 Longans, 224 Tannia, 225 Asparagus, 226 Cape Gooseberry, 237 Pitahayas, 238 Sweet Cassava, 245 Oranges, 246 Rambutan.

other damages or malformations. They also usually contain minimum requirements for other quality provisions such as maturity indicators, freshness, soundness and cleanness. For quality classification purposes, specific requirements are given for each class that is normally used in international trade for that product (e.g. "extra", class I, class II).

The Codex commodity standards normally also refer to non-product-specific Codex standards (the so-called horizontal standards) on packaging, transport and labelling, as well as maximum levels of additives and contaminants and maximum pesticide residue limits that should be observed.

UN/ECE and the OECD Scheme

The Working Party on Standardization of Perishable Produce and Quality Development of the United Nations Economic Commission for Europe (UN/ECE) has been working since 1949 on commercial quality standards for fresh fruit and vegetables, dry and dried fruit, potatoes, meat products, eggs and egg products and cut flowers.

The standard layout for UN/ECE standards for fresh fruits and vegetables contains provisions for quality (sound, clean, practically free from pest and damage by pests and free of any foreign smell and/or taste) maturity requirements and classification. Furthermore it contains provisions on sizing, tolerances, presentation and marking (=labelling). Depending on the nature of the produce a list of varieties may be included with information on known trademark protections.

There are 50 UN/ECE standards for fresh fruits and vegetables 22. http://www. unece.org/trade/agr/standard/fresh/fresh_e.htm

To avoid duplication, the Codex Committee on Fresh Fruits and Vegetables consults with UN/ECE in the elaboration of standards and codes of practice and a number of standards have been developed in Joint Codex/UNECE Groups of Experts (e.g. on fruit juices).

The OECD Scheme for the Application of Standards for Fresh Fruit and Vegetables facilitates the adaptation of quality standards to present production, trade and marketing conditions, promotes uniform quality control procedures and disseminates quality assurance guidelines. The OECD standards are the same as the UN/ECE standards.

The Scheme develops explanatory brochures and provides guidance for the application of quality assurance and inspection systems. The Scheme is open to any member country of the World Trade Organization or the United Nations or one of its Specialized Agencies. Twenty-three countries currently participate in the Scheme.

http://www.oecd.org/department/0,2688,en_2649_33907_1_1_1_1_1,00.html

International Organization for Standardization (ISO)

Quality determination, transport, storage and bar coding

ISO's Technical Committee 34 (TC 34) is responsible for developing standards for quality determinations in agricultural products (e.g. standards for the determination of acidity in fruit and vegetable products). They are basically standards for laboratories. However, Subcommittee 14 of this Technical Committee 34 has also developed guides for storage and transport for a variety of fruits and vegetables, for example ISO 931:1980 Green bananas – guide to storage and transport, or ISO 8683:1988 Lettuce – guide to pre-cooling and refrigerated transport.

---

[22] They are: FFV 02 Apricots, 03 Artichokes, 04 Asparagus, 05 Aubergines, 06 Beans, 07 Bilberries and blueberries, 08 Brussels sprout, 09 headed cabbage, 10 carrots, 11 cauliflowers, 12 ribbed celery, 13 cherries, 14 citrus fruit, 15 cucumbers, 16 fennel, 17 fresh figs, 18 garlic, 19 table grapes, 20 horse-radish, 21 leeks, 22 lettuce and endives, 23 melons, 24 cultivated mushrooms, 25 onions, 26 peaches and nectarines, 27 peas, 28 sweet peppers, 29 plums, 30 early potatoes, 31 ware potatoes, 32 raspberries, 33 scorzonera, 34 spinach, 35 strawberries, 36 tomatoes, 37 watermelons, 38 chicory (witloof), 39 edible sweet chestnuts, 40 rhubarb, 41 courgettes, 42 avocados, 43 radishes, 44 chinese cabbages, 45 mangoes, 46 kiwifruit, 47 annonas, 48 broccoli, 49 pineapples, 50 apples, 51 pears

The Codex Alimentarius Commission is in liaison with TC 34.

More recently, TC 34 working group 8 on food safety management systems developed the ISO 22000 series on food safety (see Chapter 3.1).[23]

ISO also has various standards on automatic information and data capture techniques, including bar coding. They are developed by subcommittee 31 of technical committee 1.

These include standards on terminology, symbology specifications and equipment quality and testing.

Furthermore ISO is developing the standard Traceability in the feed and food chain - General principles and basic requirements for system design and implementation. Publication is expected at the end of 2006. The draft is already available under code ISO/DIS 22005. This standard will be part of the ISO 22000 series discussed in chapter 3.1.

ISO 9000

The ISO 9000 series on quality management systems was originally published in 1994 and revised in 2000. The original ISO 9001, 9002 and 9003 were integrated into one standard: ISO 9001:2000. Now the series consists of:

ISO 9000:2000 Quality management systems – Fundamentals and vocabulary

ISO 9001:2000 Quality management systems – Requirements

ISO 9004:2000 Quality management systems – Guidelines for performance improvement

Some standards for quality management in specific situations are: ISO 10006:2003, 10007:2003 and 10012:2003.

ISO 10015:1999: Quality management – Guidelines for training

ISO 19011:2002 Guidelines for quality and/or environmental management systems auditing

ISO 9001:2000 is the only standard in the 9000 series against which a company can seek certification by an external agency. ISO 9001:2000 specifies requirements for a quality management system for any organization that needs to demonstrate its ability to consistently provide products (or services) that meet customer and applicable regulatory requirements. ISO 9001 certifications certify management systems and not products. Therefore, products cannot be labelled as ISO 9001 certified, but an indication that the firm producing the product is ISO 9001 certified is permitted (still without using the ISO logo).

ISO 9001 has been adopted globally and in many sectors has become a "default certification". The standard contains requirements for documentation and the development of a quality manual. Frequently complaints are heard that many auditors have only looked at those two aspects. Furthermore, widespread conflicts of interests have been reported, with the same company consulting about implementation as well as conducting the audits. Both developments, global implementation and non-rigorous auditing, have resulted in ISO 9000 certification losing part of its significance in the market.

2.2. NATIONAL REGULATIONS

United States

Classification

The US product codes for imports and exports are normally called schedule B numbers. The first six digits are those of the HS system, after which they will likely be very close to or the same as those of the exporting/importing country.[24]

---

[23] ISO TC34 http://www.iso.org/iso/en/CatalogueListPage.CatalogueList?COMMID=1337&scopelist=PROGRAMME
[24] Trade Information Center 1-800-USA-TRADE http://www.ita.doc.gov/td/tic/tariff/hs_numbers.htm

Quality classification/ grades

In the United States quality classifications are called grades. The USDA standards for fruits and vegetables defining these grades can be found on: http://www.ams.usda.gov/standards/stanfrfv.htm

These standards cover only products produced in the United States. However, Section 8e of the Agricultural Marketing Agreement Act of 1937 (Act) provides that when certain domestically produced commodities are regulated under a Federal marketing order, imports of the commodity must meet the same or comparable grade, size, quality and maturity requirements.[25] For these commodities, inspection of the grading and quality by the Agricultural Marketing Service of the USDA is required for each lot (shipment) imported. However, this applies only during the period of time that the domestic commodity is also being shipped and regulated, and not for counterseasonal imports. Section 8e import regulations are consistent with the purpose of Article III of the GATT. For more on Section 8e requirements, see: http://www.ams.usda.gov/fv/moab-8e.html.

The Perishable Agricultural Commodities Act (PACA)

PACA fosters trading practices in the marketing of fresh and frozen fruits and vegetables in interstate and foreign commerce. It prohibits unfair and fraudulent practices and provides a means of enforcing contracts. Under the PACA, anyone buying or selling commercial quantities of fruit and vegetables must be licensed by the USDA.

According to PACA, "Good Delivery" in connection with free on board (FOB) contracts of purchase and sale means that the commodity meets the requirements of the contract at the time of loading or sale. Moreover, if the shipment is handled under normal transportation service and conditions, it will meet the requirements on delivery at the contract destination.

In connection, the USDA has also developed FOB Good Delivery Guidelines, which specify tolerances for condition defects allowable under the assumption of a typical five-day transit time by truck (for domestic trade in the United States). http://www.ams.usda.gov/fv/paca.htm.

Food quality assurance and Commercial Item Descriptions (CIDs) for government procurement

Since 1979 the US Government-wide Food Quality Assurance Program has been assisting governmental agencies in procuring food. The programme develops Commercial Item Description (CID) formats for food items and establishes quality assurance policies and procedures applicable to the procurement of food by interested parties. CIDs concisely describe salient characteristics of commercial products. There are 139 CIDS, including for processed and fresh cut fruits and vegetables. Food purchasing agencies that use CIDs may choose to require certification. Certification may be done by USDA, in which case the USDA will select random samples of the packaged product and evaluate the conformance of the product based on the requirements of the CID. http://www.ams.usda.gov/fqa/fqfqap.htm

Country of origin labelling

On 13 May 2002, the Farm Security and Rural Investment Act of 2002, more commonly known as the 2002 Farm Bill, was signed into law. One of its many initiatives requires country of origin labelling (COOL) for beef, lamb, pork, fish, perishable agricultural commodities and peanuts. Implementation of the COOL law was foreseen for September

---

[25] Currently, these commodities are: avocados, dates (other than dates for processing), hazelnuts (filberts), grapefruit, table grapes, kiwifruit, olives (other than Spanish-style), onions, oranges, Irish potatoes, plums, prunes (suspended 1 August 2003 through 31July 2006), raisins, tomatoes and walnuts.

2004. However, it has been delayed until 30 September 2006 for all covered commodities except fish and shellfish.[26]

This means that COOL remains optional for fruits and vegetables. Voluntary origin labelling should comply with Guidelines for the Interim Voluntary Country of Origin Labelling published in 2002 and available on: http://www.ams.usda.gov/cool/ls0213. pdf. For recent information on the development of the mandatory Country of Origin Labelling, please see:http://www.ams.usda.gov/cool/.

The pending mandatory COOL law has already prompted some large wholesalers and retailers to take action. For example Merchants Distributors Inc., a large wholesaler, sent a letter to its suppliers as early as August 2003 to inform them about the pending law and require them to take preparatory steps.[27]

General labelling requirements

In the United States, labelling for foods is regulated under the Federal Food Drug and Cosmetic Act and its amendments. Food labelling is required for most prepared foods, including canned and frozen fruits and vegetables.

The law states that required label information must be conspicuously displayed and in terms that the ordinary consumer is likely to read and understand under ordinary conditions of purchase and use. Details concerning type sizes, location, etc. of required label information are contained in FDA regulations (Title 21 CFR 101), which cover the requirements of the Federal Food, Drug and Cosmetic Act and the Fair Packaging and Labelling Act.

The food labelling requirements in the regulations can be summarized as follows. With a few exceptions, if the label of a food bears representations in a foreign language, the label must bear all of the required statements in the foreign language as well as in English. If the food is packaged, the following statements must appear on the label in the English language: The name and address of the manufacturer, packer or distributor. If the food is not manufactured by the person or company whose name appears on the label, the name must be qualified by "Manufactured for," "Distributed by," or similar expression. An accurate statement of the net amount of food in the package must be present. The common or usual name of a food must appear. The ingredients in a food must be listed by their common or usual names in decreasing order of their predominance by weight. Food additives and colours are required to be listed as ingredients.

Nutrition labelling, i.e. information on the nutritional value of the product, on packaged food is compulsory. Nutrition labelling for raw produce (fresh fruits and vegetables) and fish is voluntary, but should follow the FDA voluntary nutrition labelling regulations. An amendment of these voluntary labelling regulations is expected in 2005 (comment period closed June 2005). For details on the labelling requirements, please see http://www.cfsan.fda.gov/~dms/qa-indlq.html.[28] (The United States has recently notified the TBT Committee that it is considering a revision of its nutrition labelling rules in view of the increasing obesity problems in the country.)

Bioterrorism Act, prior notice and registration of food facilities

The Public Health Security and Bioterrorism Preparedness and Response Act of 2002 (the Bioterrorism Act) includes new authority for the Secretary of Health and Human Services (HHS) to take action to protect the nation's food supply against the threat of intentional contamination. The FDA, as the food regulatory arm of HHS, is responsible for developing and implementing these food safety measures. For more information on the Bioterrorism Act, see http://www.access.fda.gov and http://www.cfsan.fda.gov/~dms/fsbtact5.html.

---

[26] http://www.ams.usda.gov/cool/
[27] http://www.merchantsdistributors.com/
[28] http://www.cfsan.fda.gov/~dms/qa-indlq.html

One of the measures implemented in the wake of the Bioterrorism Act is the prior notice requirement. The FDA must be notified electronically in advance of any shipments of food that are imported into the United States. The measure has applied since December 2003 and is explained in the booklet What you need to know about prior notice of imported food shipment" available at http://www.cfsan.fda.gov/~acrobat/fsbtpn.pdf.

Another measure implemented at the same time was the requirement for registration of food facilities. All domestic and foreign facilities that manufacture, process, pack or hold food for consumption in the United States must register with the FDA. If a foreign facility sends the food to another foreign facility for further manufacturing/processing or packaging before the food is exported to the United States, only the second foreign facility is required to register. However, if the second facility performs only a minimal activity, such as putting on a label, both facilities must register. Farms are exempt from registration.

For details please see the booklet What you need to know about Registration of Food facilities at http://www.cfsan.fda.gov/~acrobat/fsbtreg.pdf.

Traceback/Traceability

Another result of the Bioterrorism Act was that traceback rules in the United States were tightened. Section 306 on administration and record keeping regulation requires that each company in the supply chain keep information about: the company from which it received the products (previous source nontransporter); the company that delivered the product (previous source transporter), the company who took it away (subsequent source transporter); and the company the product was given (sold) to (subsequent source nontransporter). The recordkeeping regulations specify what information must be made available. However, they do not specify how the records must be kept, as long as the information can be retrieved within a 24-hour period.

The Final Rule on the Establishment and Maintenance of Records to Enhance the Security of the US Food Supply Under the Bioterrorism Act was published in 2004 and is available at http://www.cfsan.fda.gov/~acrobat/fr04d09a.pdf. The final rule excludes foreign entities except for persons who transport food in the United States. For more information, see http://www.cfsan.fda.gov/~dms/fsbtac25.html. and the booklet What you need to know about establishment and maintenance of records at http://www.cfsan.fda.gov/~acrobat/fsbtrec.pdf.

European Union

For an overview of all agricultural import regulations in the European Union, see the Food and Agricultural Import Regulations and Standards (FAIRS) report, produced by the US mission to the European Union in Brussels, which is updated yearly. The issue of August 2005 is available at: http://www.fas.usda.gov/gainfiles/200508/146130611.pdf. For new editions and reports for specific EU Member States, see: http://www.useu.be/agri/fairs.html.

In the paragraphs that follow, the general regulations that are most relevant for fresh fruits and vegetables are discussed.

Classification

The European Union uses the Combined Nomenclature (CN) for the customs classification of goods. The first six digits refer to the Harmonized System codes and the two digits that follow represent the CN subheadings. Commodity codes and import duties can be found in the European Union's customs database: http://europa.eu.int/comm/taxation_customs/dds/en/tarhome.htm.

The United Economic Commission for Europe (UN/ECE) Working Party on Facilitation of International Trade Procedures is responsible for the development of standard international trade data terminology and a uniform system for automatic processing and transmission of trade data. In 1990, CEN created a technical committee on bar coding (TC 225) which has published various standards on terminology, symbology specifications and equipment quality and testing.

http://www.cenorm.be/CENORM/BusinessDomains/TechnicalCommitteesWorkshops/CENTechnicalCommittees/Standards.asp?param=6206&title=CEN%2FTC+225

Quality classification

Since January 2002, all imports of fruits and vegetables must have a certificate of conformity to the EU marketing standards (regulation 1148/2001). In general, these marketing standards contain specifications for the minimum quality, classification (Extra, I and II), grading and labelling. EU marketing standards normally follow the Codex commodity standards quite closely. A standard does not exist for every tropical fruit or vegetable (e.g. no standard for pineapples). For an overview of EU marketing standards see: http://www.defra.gov.uk/hort/hmi.htm.

Each importer has to apply for a certificate to the relevant authority in the country of import. The procedures and inspections are organized differently in each country. A list of coordinating authorities designated by the Member States for the purpose of conformity checks in the sector of fruit and vegetables is available at: http://europa.eu.int/eurlex/lex/LexUriServ/site/en/oj/2005/c_023/c_02320050128en00090013.pdf

In April 2004 Regulation 907/2004 amended the marketing standards with regard to the presentation and labelling provisions. The amendment was in response to the increased use of pre-packaging and reusable transport packages. In the original regulation all packages had to include the name and address of the packer/dispatcher, the nature of the produce, its origin and commercial specification. Now, transport packages are exempt from this requirement provided that the sales packages are correctly labelled and visible from the outside. The amendment also adds that stickers on individual products may not leave visible traces or glue and may not damage the skin of the fruit after they have been removed.

Import licenses and the entry price system

Fresh fruits and vegetables are not generally subject to import licence restrictions. For garlic, a quota system is in place, which is monitored by the issuing of licenses. The banana import regime has recently been changed into a tariff-only system.

However, 15 products are subject to an entry price system.[29] This system sets a minimum import value for each product depending on the EU season. If the CIF import value of a consignment is below the entry price, but not more than 8 percent below, an additional duty is charged which equals the difference between the entry price and the import price. If the import price is more than 8 percent below the entry price, the full maximum tariff equivalent duty is charged in addition to normal customs duty (ad valorem duty).

The system is complicated by special agreements, such as those with Mediterranean countries under regulation 747/2001. This regulation sets quotas for specific products for each country. Imports within the quota are exempt from the normal customs duty. For products with an entry price system, the system applies also within the quota.

The provision for the entry price system and tariff quotas resulting from agreements within the Framework of the Uruguay Round is stipulated in Council Regulation (EC) No 2200/96 of 28 October 1996 on the common organization of the market in fruit and vegetables. The regulation and all amendments and corrigenda can be downloaded from:

http://europa.eu.int/smartapi/cgi/sga_doc?smartapi!celexapi!prod!CELEXnumdo

---

[29] These are: apples, apricots, artichokes, cherries, clementines, courgettes, cucumbers, grapes, lemons, mandarins, oranges, peach/nectarines, pears, plums and tomatoes.

c&lg=EN&numdoc=31996R2200&model=guicheti (Note: the site does not provide information on actual duties or entry prices)

The tariff quota system for garlic can be found at:

http://europa.eu.int/smartapi/cgi/sga_doc?smartapi!celexapi!prod!CELEXnumdoc&lg=EN&numdoc=32002R0565&model=guicheti

The consolidated text (January 2004) of regulation 747/2001 for Mediterranean countries can be found at: http://europa.eu.int/eur-lex/en/consleg/pdf/2001/en_2001R0747_do_001.pdf.

For subsequent amendments see:

http://europa.eu.int/smartapi/cgi/sga_doc?smartapi!celexapi!prod!CELEXnumdoc&lg=EN&numdoc=32001R0747&model=guicheti

Traceback/traceability

In January 2005 article 18 of regulation EC 178/2002 (the General Food Law, establishing the European Food Safety Authority) came into effect. This article stipulated that "the traceability of food [...] shall be established at all stages of production, processing and distribution."

The European Union published a guidance document in December 2004. The document explains that article 18 requires food business operators inside the European Union to identify from whom and to whom a food product or ingredient has been supplied (the "one step back, one step forward" approach). The traceability provisions do not have an extraterritorial effect outside the European Union. It is enough that the European Union importer is able to identify from whom the product was exported in the third country. The term "supply" should not be interpreted as mere delivery, so the identification of the shipper does not fulfil the requirements.

Article 18 does not specify what information should be kept, but the guidance document considers the following necessary to fulfil the objective of article 18:

- name and address of supplier, nature of products which were supplied.
- name and address of customer, nature of products that were delivered to that customer.
- date of transaction/delivery.

Furthermore, it is highly recommended that information be kept on the volume and batch number, and a more detailed description of the product (e.g. variety of fruit/vegetable). It is essential that the traceability system be designed to follow the physical flow and not only the invoices. Article 18 does not foresee a minimum period of time for recordkeeping, but the guidelines indicate a period of six months for highly perishable products such as fruits and vegetables. http://europa.eu.int/comm/food/food/foodlaw/guidance/index_en.htm.

Implementation of food law and third-country status

Regulation 882/2004 sets rules for the official controls and actions to be undertaken to ensure compliance with the food law. The regulation will be applicable as of 1 January 2006. Frequency of physical checks will depend on: risks associated with the type of food; the history of compliance (of the country of origin and the operators); the controls carried out by the importer; and the guarantees given by the competent authority of the third country of origin.

Approved third countries may inspect and issue certificates at the port of departure. These shipments may still be subject to spot checks. Approved third countries (non EU-25) are (as per end 2003) India, Israel, Morocco, South Africa and Switzerland. These inspections and certifications may cover compliance with the food law as well as with the marketing standards. For details on the implementation of the food law and control system, see regulation 882/2004 (cor.):

http://europa.eu.int/eur-lex/pri/en/oj/dat/2004/l_191/l_19120040528en00010052.pdf

General labelling requirements and label of origin

Labelling requirements for final sale are based on the marketing standards (1148/2001, discussed above in the paragraph on quality classification), the mixed product packaging regulation (regulation 48/2003), the lot identification directive (89-396) and the labelling directive (2000-13). The labelling directive basically requires that labelling not be misleading and that information be provided on: the name of the product; ingredients and their quantity (except for e.g. whole fruits and vegetables); (net) quantity; use by date; special storage or use conditions and instructions; name and address of manufacturer, packager or seller; and place of origin if omission may mislead the consumer. Unlike in the United States, nutrition labelling is not compulsory in the European Union, unless any nutrition claims are made on the label or in advertising. See http://europa.eu.int/scadplus/leg/en/lvb/l21092.htm.

2.3. CORPORATE STANDARDS

Bar codes: the Universal Product Code (UPC) and Global Standards 1/European Article Numbering-Uniform Code Council (GS1/EAN-) systems

The UPC system for bar codes was developed in the United States by a grocery trade association. Directly upon acceptance of the numeric system, the optical bar code symbol was developed which contains the numeric information. The bar code originally had 12 digits. On request of the EAN system, in the 1970s a thirteenth digit was added for country specification. In practice, the UPC system is now a subset of the EAN- system.

Membership to EAN was extended to organizations from outside Europe, and the name was changed to EAN International in 1992 and again to GS1. Today GS1 has 101 member organizations representing companies in 103 countries. Member organizations are usually national associations of companies. They allocate unique company numbers (company prefix) to their members and manage the allocation of Global Location Numbers, Global Trade Item Numbers and Serial Shipping Container Codes. They also provide training on numbering, bar coding and Electronic Data Interchange (EDI).

To be able to use the EAN- system, a company number is needed, and hence membership in GS1. Membership fees differ between membership organizations. Companies in countries without a member organization can apply for direct GS1 individual membership.

Traceability guidelines for the North American fresh produce industry

The Canadian Produce Marketing Association (CPMA) and the Produce Marketing Association (PMA) from the United States joined forces in late 2002 in the CPMA/PMA Traceability Task Force. Members include 17 appointed broad-based representatives from the North American produce industry, consisting of grower-shippers, retailers, food service operators, wholesalers, distributors and regional produce associations.
They produced a collection of best-practice examples in traceability, available at
http://www.pma.com/Template.cfm?Section=Traceability&CONTENTID=4662&TEMPLAT E=/ContentManagement/ContentDisplay.cfm.

Subsequently they produced a guide for implementing traceability systems (March 2005):

http://www.pma.com/Template.cfm?Section=Traceability&CONTENTID=7399&TE MPLATE=/ContentManagement/ContentDisplay.cfm. Whereas the United States and European Union require segmented traceability (one step up, one step down), they encourage whole-chain traceability. They give guidance for paper-based, bar-code based and RFID-based systems.

Furthermore, they endorse the Can-Trace Data Standard as the North American produce traceability data standard. This Standard defines the data requirements but does not advise which data carriers to use. The Can-Trace Standard (version 1.0) was published in November 2004. A stakeholder consultation process was organized in the spring of 2005, and on the basis of the comments, a second version was produced. The standard can be downloaded from:

http://www.can-trace.org/portals/0/docs/CFTDS%20version%202.0%20FINAL.pdf

Qualipom'fel

In France, a certification programme for fruit and vegetable wholesalers, Qualipom'fel, has been in place since 1995. The standard was developed in a tripartite structure with government, the industry (represented by l'Union) and its clients. The standard is organized in seven engagements/promises/obligations. One of these regards the relations with the suppliers. The wholesalers are supposed to communicate quality requirements to them on the basis of formalized criteria, evaluate the suppliers' attitude versus these criteria, and check incoming produce. For wholesalers who used to have only telephone contact with their suppliers to agree on volume and price, this was a substantial change in the way of doing business.

http://www.qualipomfel.com/index.htm

L'Union, the industry association of fruit and vegetable wholesalers in France, also developed another quality initiative: Fel'engagement. This is a procedure for quality monitoring of the produce at the point of reception by the wholesaler. Participants receive recognition by the Directorate of Competition, Consumption and Repression of Fraud (DGCCRF). http://www.qualipomfel.com/felengagement.html

3. Phytosanitary standards

A useful source of information on phytosanitary regulations is the International Portal on Food Safety, Animal Health and Plant Health (http://www.ipfsaph.org/En/default. jsp). This reader-friendly portal includes IPPC standards, WTO rules and notifications and national regulations.

3.1. INTERNATIONAL AGREEMENTS

As discussed in chapter 1.1, the WTO SPS Agreement is not a standard, but a framework for standard setting around sanitary and phytosanitary issues. It recognizes the need for SPS standards but at the same time tries to prevent SPS standards from being used for trade protectionism. To this end, the SPS Agreement promotes the use of internationally harmonized standards. In the area of phytosanitary standards, it specifically mentions the International Plant Protection Convention (IPPC).

International Plant Protection Convention (IPPC)

The IPPC is an international treaty to prevent the introduction and spread of plant and plant product pests, and to promote appropriate measures for their control. The convention was initially adopted by the Conference of FAO in 1951. Currently (as of 25 April 2005) 137 governments adhere to the convention. The IPPC Secretariat coordinates the activities of the Convention and is hosted by FAO.

The Convention was last revised in 1997, and the new text will come into force after it is accepted by two-thirds of the contracting parties to the IPPC. As of 21 June 2005, there were still nine signatures needed. The revision provides for the establishment of a Commission on Phytosanitary Measures to be the standard-setting body for International Standards for Phytosanitary Measures (ISPMs). An Interim Commission on Phytosanitary Measures has been established by FAO until the New Revised Text comes into force.

The IPPC is a legally binding international agreement, but the standards developed and adopted by the Convention are not legally binding under the IPPC. However, WTO members are required to base their phytosanitary measures on international standards developed within the framework of the IPPC. Currently there are 24 adopted ISPMs. They cover principles for: plant quarantine; determination of pest status and pest risk analysis; the use of biological control agents; the establishment of pest free areas; export certification systems; etc.

The Convention itself requires its contracting parties to issue phytosanitary certificates of compliance with the phytosanitary regulations of other contracting parties. ISPM 7 describes an export certification system to produce valid and credible phytosanitary certificates.

ISPM 15 on wood packaging is being used by increasingly more countries to prevent the introduction of wood pests. For export to these countries, wooden packaging must be pre-treated in a specifically prescribed way.

More information on the IPPC and the texts of adopted and draft standards can be found on: https://www.ippc.int/IPP/En/default.jsp

3.2. NATIONAL REGULATIONS
United States
Import permits

Plant quarantine regulations for phytosanitary reasons are divided into two classes – prohibitory and restrictive. Prohibitory orders forbid the entry of designated plants and plant products that are subject to attack by plant pests or plant diseases for which there is no treatment available. Restrictive orders allow the entry of plants or plant products under either a treatment or an inspection requirement.

The Animal and Plant Health Inspection Service (APHIS) is responsible for implementing phytosanitary controls and measures. For most fruits and vegetables, import permits are required as per federal regulation Title 7 CFR 319.56 Foreign Quarantine Notices, subpart Fruits and vegetables. http://199.132.50.50/Oxygen_FOD/FB_MD_PPQ.nsf/d259f66c6afbd45e852568a90027bcad/4fd5f501221cc9db852568f2006b255a/$FILE/0018.pdf

The manual Regulating the Importation of Fresh Fruit and Vegetables is intended for the inspection services, but it is also very useful for exporters, since it includes lists of approved fruits and vegetables for import per country of origin. The latest update can be downloaded from

http://www.aphis.usda.gov/ppq/manuals/pdf_files/FV_Chapters.htm.

The permit application process is quite simple for these approved fruits and vegetables. Permits are free and valid for five years so long as the pest/disease risk has not changed and/or so long as actual shipments are not found violative. However, achieving admissibility for a new product from a foreign country can be time consuming and costly.

An explanation for obtaining permits is given at: http://www.aphis.usda.gov/ppq/permits

The Aphis Import Authorization System currently allows customers to submit applications for import permits online. https://web01.aphis.usda.gov/IAS.nsf/Mainform?OpenForm

Aphis has developed pre-clearance programmes with plant protection services and export industries in exporting countries, but its web site does not indicate which countries or industries.

http://www.aphis.usda.gov/ppq/preclearance/

Bioterrorism Act

The Agricultural Bioterrorism Protection Act of 2002 also has some phytosanitary provisions. A list of selected agents and toxins deemed to pose a severe threat to animal or plant health or products was published in March 2005. Entities handling such agents and toxins must register themselves.

Wood packaging

APHIS has set new standards for Wood Packaging Material imported into the United States through Title 7 CFR 319.40 – Importation of Wood Packaging Material, published on 16 September 2004. This rule states that all regulated wood packaging material shall be appropriately treated and marked under an official programme developed and overseen by the National Plant Protection Organization (NPPO) in the country of export. This new regulation has been gradually implemented from September 2005 and full enforcement is planned to start from July 2006. Countries are also required to comply with ISPM 15. More information, including the rule, is available at:www.aphis.usda.gov/ppq/wpm/import.html.

Heat treatment

In 1984 the United States prohibited fumigation by ethylene dibromide because it had been found to be carcinogenic. However, for several fruit flies the USDA security level of quarantine pest control (no more than 3.2 survivors out of 100 000 larvae) still needed to be met. Several heat treatment methods for mango and other susceptible fruit were tested. Hot water immersions are currently used for mango. There are now approximately 85 commercial hot water treatment facilities in Latin America. The cost of core equipment is only around US$200 000, but additional cold rooms, water purification equipment and other refinements can bring the total investment to as high as US$ 1 million. APHIS/PPQ must certify the facility, and each treated batch and truckload must be checked by an approved inspector.[30]

Europe
European and Mediterranean Plant Protection Organization (EPPO)
EPPO is an intergovernmental organization of 47 member countries responsible for cooperation in plant protection in the European and Mediterranean region. Under the International Plant Protection Convention (IPPC), EPPO is the regional plant protection organization for Europe.

EPPO makes recommendations (in the form of Regional Standards) to the National Plant Protection Organizations of its member countries. These recommendations are Regional Standards in the sense of the revised IPPC. EPPO Standards have been developed for plant protection products and for phytosanitary measures.

http://www.eppo.org/Standards/standards.html#pms

EU phytosanitary certification for fruits and vegetables
For imports into the European Union, certain plants, plant products or other objects (listed in Part B of Annex V to Directive 2000/29/EC) must be accompanied by a phytosanitary certificate, issued by the National Plant Protection Organization of the exporting country. These include various fruits and vegetables intended for commercial sale.[31]

Phytosanitary certificates should be issued conforming to the models set out under the IPPC, certifying that the plants, plant products or other objects: have been subject to the appropriate inspections; are considered to be free from quarantine harmful organisms and practically free from other harmful organisms; and are considered to conform to the phytosanitary regulations of the importing country. All Member States are required to introduce the harmonized phytosanitary certificates by 1 July 2005.[32]

As of 1 January 2005, each individual consignment of controlled plants and produce that requires a certificate must undergo an identity check and physical plant health inspection. However, a reduction in the frequency of inspections is allowed if the number of consignments from the same origin that was found to be infected by harmful organisms is less than 1 percent of at least 600 consignments.(For details, please see Commission Regulation 1756/2004.

http://europa.eu.int/eur-lex/pri/en/oj/dat/2004/l_313/l_31320041012en00060009.pdf).

The EU Commission has agreed that from 1 January 2005, reduced levels of inspection can be carried out on 32 trades. Reduced levels of inspection range from 3 to 50 percent. (For more information, see http://www.defra.gov.uk/planth/newsitems/reduced.pdf). For many countries of origin the new inspection regime means an increase in the number of physical inspections.

---

[30] Lamb, J.E., Velez J.A. and Barclay R.W. 2004. The Challenge of Compliance with SPS and Other Standards Associated with the Export of Shrimp and Selected Fresh Produce Items to the United States Market. World Bank Agriculture and Rural development Discussion paper. (page 41)

[31] These are (as of May 2004): citrus fruits (including Kumquat), Momordica (e.g. balsampear and bitter gourd), eggplant, Annona (e.g. cherimoya, soursop, durian, custard apple), quince, Diospyros (Persimmon, sapote), apple, mango, passion fruit, prunus (plums, prunes, cherries and almonds), guava, pear, ribes (e.g. currant), syzygium (riberry, rose apple), vaccinium (blueberry, cranberry) and grapes. See the consolidated version of Directive 2000/29 and its amendments of May 2004: http://europa.eu.int/eur-lex/en/consleg/pdf/2000/en_2000L0029_do_001.pdf

[32] Summary report of the meeting of the standing committee on plant health (SCPH) held on 26-27 May 2005 [05/05] http://europa.eu.int/comm/food/fs/rc/scph/rap81_en.pdf

Wood packaging

The European Union has adapted its directive on solid wood packing material according to ISPM-15 (Directive 2004/102/EC). The new provisions came into effect in March 2005 for wood thicker than 6mm. The measures cover all packing material made of non-manufactured coniferous wood, including packing cases, boxes, crates, drums, pallets, box pallets and other load boards. Pallets must satisfy the standard of the UIC (International Union of Railways) and be marked accordingly. From March 2006, in addition to ISPM-15 requirements, solid wood packaging material is required to be debarked. For details, please see the directive at
   http://europa.eu.int/eurlex/pri/en/oj/dat/2004/l_309/l_30920041006en00090025.pdf.

4. Food safety standards

A useful source of information on food safety regulations is the International Portal on Food Safety, Animal Health and Plant Health (http://www.ipfsaph.org/En/default.jsp). This user-friendly portal includes Codex standards, WTO rules and notifications and national regulations.

4.1. INTERNATIONAL AGREEMENTS

Codex

General features of Codex were outlined under Chapter 1.1. The commodity quality standards, which generally also contain food safety-related requirements, were discussed under chapter 2.1. This paragraph concentrates on food safety standards that are valid across commodities.

Food hygiene

CAC/RCP 1-1969, Rev. 4-2003. Recommended International Code of Practice. General Principles of Food Hygiene.

These are basic rules for the hygienic handling, storage, processing, distribution and final preparation of all food along the food production chain. This document should serve as a basis for the establishment of Good Hygiene Practices (GHP). Topics addressed by the code include: design and adequate facilities; control of operations (including temperature, raw materials, water supply, documentation and recall procedures); maintenance and sanitation; personal hygiene; and training of personnel. This code contains the Annex on Hazard Analysis and Critical Control Point (HACCP) System and Guidelines for its Application.

www.codexalimentarius.net/download/standards/23/CXC_001_2003e.pdf

CAC/GL 21-1997. Principles for the Establishment and Application of Microbiological Criteria for Foods www.codexalimentarius.net/download/standards/394/CXG_021e.pdf

CAC/GL 30-1999. Principles and Guidelines for the Conduct of Microbiological Risk Assessment www.codexalimentarius.net/download/standards/357/CXG_030e.pdf

CAC/RCP 53-2003. Code of Hygienic Practice for Fresh Fruits and Vegetables

This code of practice covers general hygienic practices for the primary production and packing of fresh fruits and vegetables cultivated for human consumption. The code has been established in order to produce a safe and wholesome product, particularly for those products intended to be consumed raw. The code is applicable to fresh fruits and vegetables grown in the field (with or without cover) or in protected facilities (hydroponic systems, greenhouses). It concentrates on microbial hazards and addresses physical and chemical hazards only insofar as these relate to good agricultural practices (GAP) and good manufacturing practices (GMP). The code does not provide recommendations for handling practices to maintain safety of fresh fruits and vegetables at wholesale, retail, food services or in the home.

This code is supplemented by the Annex for Ready-to-eat Fresh Pre-cut Fruits and Vegetables (Annex I) and the Annex for Sprout Production (Annex II). Annex I covers the hygienic practices for the processing of ready-to-eat, pre-cut fruits and vegetables. Annex II covers the hygienic practices that are specific to the primary production of seeds for sprouting and the production of sprouts for human consumption.

Hazard Analysis Critical Control Point (HACCP)

The HACCP system is a widely accepted food safety management system. It originated in the 1960s in the United States for aerospace purposes. In 1985, the United States National Academy of Science recommended that the HACCP approach be adopted in food processing establishments to ensure food safety. Currently, many international organizations recommend HACCP and many national food safety regulations and corporate standards require its implementation.

In 1993 the Codex Alimentarius Commission adopted Guidelines for the application of the HACCP system (ALINORM 93/13A, Appendix II). In 1997 the HACCP guidelines were incorporated into the food hygiene code (CAC/RCP 1, see above) as an Annex.

Codex definitions:

HACCP: A system which identifies, evaluates and controls hazards that are significant for food safety.

- Hazard: A biological, chemical or physical agent in, or condition of, food with the potential to cause an adverse health effect.
- Hazard analysis: The process of collecting and evaluating information on hazards and conditions leading to their presence to decide which are significant for food safety and therefore should be addressed in the HACCP plan.
- Critical Control Point: A step at which control can be applied and is essential to prevent or eliminate a food safety hazard or reduce it to an acceptable level.
- Control measure: Any action and activity that can be used to prevent or eliminate a food safety hazard or reduce it to an acceptable level.

The HACCP system consists of 12 tasks, related to 7 principles:

- assemble the HACCP team
- describe product
- identify intended use
- construct flow diagram (and plant schematic)
- on-site confirmation of flow diagram (and plant schematic)
- list all potential hazards associated with each step, conduct a hazard analysis and consider any measures to control identified hazards - Task 6/Principle 1
- determine critical control points - Task 7/Principle 2
- establish critical limits for each critical control point - Task 8/Principle 3
- establish a monitoring system for each critical control point - Task 9/Principle 4
- establish corrective actions - Task 10/Principle 5
- establish verification procedures - Task 11/Principle 6
- establish documentation and record keeping - Task 12/Principle 7

In other words, the HACCP system is designed to control significant hazards at those points in the food chain where its control is most effective and efficient (the critical control point).

Pesticide residues

The Codex Committee on Pesticide Residues (CCPR) is responsible for identifying substances that require priority evaluation. The evaluation is then carried out by the independent Joint FAO/WHO Meeting on Pesticide Residues (JMPR). Based on the JMPR evaluation, the CCPR prepares draft standards for submission to the Codex Alimentarius Commission (CAC).

There are currently (2005) 2 579 maximum limits for pesticide residues, covering 213 pesticides. Codex maximum residue limits for pesticides can be most easily consulted via FAOSTAT:

http://faostat.fao.org/faostat/collections?version=int&hasbulk=1&subset=FoodQuality

Additives and contaminants

Similar to the procedure for pesticide residue limits, the Codex Committee on Food Additives and Contaminants (CCFAC) identifies substances for priority evaluation. Substances are then evaluated by the independent Joint FAO/WHO Expert Committee on Food Additives (JECFA). Based on the evaluation, the CCFAC prepares maximum levels for additives and contaminants for inclusion in the corresponding general Standards for Food Additives, Contaminants and Toxins in Foods after adoption by the Commission.

CX STAN 192-1995, Rev.5-2004.
Codex General Standard for Food Additives.

There are currently (2005) 638 food additive provisions covering 222 food additives. In relation to fresh fruits and vegetables, there are a number of adopted maximum levels for additives, mainly waxes. However, for fresh fruits and vegetables additives are not very relevant.

Codex texts for contaminants include maximum and guideline levels for contaminants as well as codes of practices for a number of contaminants-food combinations. Selected relevant texts for fruits and vegetables are cited below:

CX STAN 193-1995, Rev.1-1997.
Codex General Standard for Contaminants and Toxins in Foods.
www.codexalimentarius.net/download/standards/17/CXS_193_2004e.pdf

CAC/RCP 49-2001.
Code of Practice for source directed measures to reduce contamination of food with chemicals.
www.codexalimentarius.net/download/standards/373/CXP_049e.pdf

CAC/RCP 56-2004.
Code of Practice for the prevention and reduction of lead contamination in foods.
www.codexalimentarius.net/download/standards/10099/CXC_056_2004e.pdf

Irradiation
Irradiation is increasingly used to be able to preserve fresh fruits and vegetables longer, without changing the appearance and taste of the product.

CX STAN 106-1983, Rev.1-2003.
Codex general standard for irradiated food.

This standard applies to foods processed by ionizing radiation. This is used in conjunction with applicable codes of practice and food standards including conditions for re-irradiation of foods. It does not apply to foods exposed to doses imparted by measuring instruments used for inspection purposes.

In addition to mandatory labelling requirements laid down in the General Standard for the Labelling of Pre-packaged Foods, this standard states that the declaration of the fact of irradiation should be made clear on the relevant shipping documents. In the case of products sold in bulk to the consumer, the international logo and the words "irradiated" or "treated with ionizing radiation" should appear, together with the name of the product on the container in which products are placed.
www.codexalimentarius.net/download/standards/16/CXS_106_2003e.pdf

Foods Derived from Biotechnology
The following texts address the risk analysis and risk assessment aspects of foods derived from recombinant-DNA organisms.

CAC/GL 44-2003

Codex Principles for Risk Analysis of Foods derived from Modern Biotechnology.

An overarching framework for undertaking risk analysis on the safety and nutritional aspect of foods derived from biotechnology.

CAC/GL 45-2003

Codex Guideline for the Conduct of Food Safety Assessment of Foods Derived from Recombinant-DNA Plants

A guideline, based on the above principles, for conducting safety assessments specifically for foods derived from recombinant-DNA plants

CAC/GL 46-2003

Codex Guideline for the Conduct of Food Safety Assessment of Foods Derived from Recombinant-DNA Micro-organisms

A guideline, based on the principles, for conducting safety assessments specifically for foods derived from recombinant-DNA micro-organisms.

ISO 22000

ISO published its new standard for food safety management systems, ISO 22000, in September 2005. ISO 22000 specifies the requirements for a food safety management system in the food chain in order to provide consistently safe end products that meet both the requirements agreed with the customer and those of applicable food safety regulations.

According to ISO, ISO 22000 will combine the HACCP principles with prerequisite programmes (i.e. a specified procedure or instruction). The standard will distinguish between two types of prerequisite programmes: infrastructure & maintenance programmes and operational programmes.

According to ISO, ISO 22000 can be applied by organizations ranging from primary producers, to food manufacturers, transport and storage operators and subcontractors, to retail and food service outlets – together with interrelated organizations such as producers of equipment, packaging material, cleaning agents, additives and ingredients.

The standard can be applied on its own, or in combination with other management system standards, such as ISO 9001:2000, with or without independent (third party) certification of conformity. The publication of ISO 22000 is complemented by an ISO Technical Specification (ISO/TS 22004, published November 2005), which gives guidance on the implementation of the standard, with specific attention to small and medium-sized enterprises. In 2006, another Technical Specification ((ISO/TS 22003) will be published explaining certification requirements applicable when third-party certification is used.

These documents are being developed by working group WG 8, Food safety management systems, of ISO technical committee ISO/TC 34, Food products. Experts from 23 countries are participating, and organizations with liaison status include: Confederation of the Food and Drink Industries of the European Union (CIAA); International Hotel and Restaurant Association; CIES/Global Food Safety Initiative; and World Food Safety Organisation (WFSO).

4.2. NATIONAL FOOD SAFETY STANDARDS

United States

Food code and liability

Retail establishments that sell food must comply with the Food Code (2001 edition). The code simply states that food sold in the retail establishment must be safe. Regarding the source of the food, it simply says that the source should comply with the food law. The code includes an Annex with HACCP guidelines. http://vm.cfsan.fda.gov/%7Edms/fc01-toc.html

Hygiene and microbiological contamination

The document Guide to Minimize Microbial Food Safety Hazards for Fresh Fruits and Vegetables (1998) focuses on microbial food safety hazards for fresh fruit. It also provides good agricultural practices (GAP) and good manufacturing practices (GMP) common to the growing, harvesting, washing, sorting, packing and transporting of most fruits and vegetables sold to consumers in an unprocessed or minimally processed (raw) form. The guide does not specifically address other areas of concern to the food supply or the environment (such as pesticide residues or chemical contaminants).

The guide discusses the microbial hazards and control of potential hazards on the following subjects: water; manure and municipal biosolids; worker health and hygiene; sanitary facilities; field sanitation; packing facility sanitation; transportation; and traceback.

Although voluntary from a regulatory point of view, some US supermarket chains (e.g. Safeway) have required implementation of these FDA guidelines and third-party auditing of implementation as a prior condition of doing business.

Furthermore, the guide cites some specific federal regulations that are mandatory for domestic producers. These are:

- under the Federal Food, Drug, and Cosmetic Act and the Fair Packaging and Labelling Act (Title 21 CFR):
  - 21 CFR 110.10 on worker health and hygienic practices within the context of GMP in the manufacturing, packing or holding of human food.
  - 21 CFR 110.20 to 110.93 on good manufacturing practices for buildings and facilities, equipment, and production and process controls for foods, including rules on water used for food and food contact surfaces in processing facilities.
  - 21 CFR 179 on irradiation (see paragraph on irradiation below).
- under the Occupational Safety and Health Act (Title 29 CFR):
  - 29 CFR 1928.110 subpart I, on field sanitation.
  - 29 CFR 1910.141, subpart J, on sanitation for enclosed packing facilities.

The guide may be downloaded from: http://www.foodsafety.gov/~acrobat/prodguid.pdf

More recently the Food and Drug Administration has developed a food safety action plan that looks to the entire produce chain: Produce Safety from Production to Consumption: 2004 Action Plan to Minimize Foodborne Illness Associated with Fresh Produce Consumption. June, 2004. The action plan has four objectives: prevent contamination of fresh produce; minimize the public health impact when contamination of fresh produce occurs; improve communication with producers, preparers and consumers about fresh produce; and facilitate and support research relevant to fresh produce. http://www.cfsan.fda.gov/~dms/prodpla2.html

Pesticide residues

In the United States, the Environmental Protection Agency (EPA) is responsible for establishing maximum residue limits, which are known as tolerances. Since 1998 EPA has been conducting a re-registration process for older agrochemicals, and since the Food Quality Protection Act of 1996 pesticides must meet new food safety standards. As a consequence, many agrochemicals viewed as necessary for imported specialty products have still not been re-registered. The Pesticide Registration Improvement Act, which came into effect in March 2004, is meant to speed up the process and make it more predictable.

The tolerance database previously offered by EPA is being reviewed and updated. In the interim, tolerance information can be found within Title 40 CFR 180, as published 1 July 2004: http://www.access.gpo.gov/nara/cfr/waisidx_04/40cfr180_04.html.

For information on new tolerances or changes to tolerances since 1 July 2004, see EPA's Federal Register site: http://www.epa.gov/fedrgstr/EPA-PEST/index.html.

When the EPA database is updated it will again be available at:

http://www.epa.gov/pesticides/food/viewtols.htm

Another source of information on residue tolerances is the National Pesticide Information Retrieval System (NPIRS). The NPIRS is a collection of pesticide-related databases available by subscription. http://aboutnpirs.ceris.purdue.edu/

Irradiation

Irradiation is regulated by Federal Regulation 21 CFR 179 Irradiation in the Production, Processing and Handling of Food. This regulation establishes maximum (and sometimes minimum) doses for specific uses. Food that has been irradiated must be labelled with the radura symbol accompanied by the claim "treated by irradiation" or "treated with radiation".[33]

http://www.access.gpo.gov/nara/cfr/waisidx_02/21cfr179_02.html

Genetically Modified Foods

Genetically modified crops must be approved before they can be grown commercially in the United States. There are no restrictions on marketing foods derived from genetically modified organisms. One of the largest food industry associations, the Food Marketing Institute (FMI), is calling for some regulation. They have asked the FDA to communicate a clear definition as to what constitutes genetically modified foods or food or food products and establish criteria for GM Free and Non-GM ingredient labelling.[34]

European Union

An exhaustive, user-friendly and up-to-date overview of all EU and Norwegian legislation related to food safety and food labelling can be found on the CBI web site. The site is regularly updated. The site can be searched for specific sectors and export destination countries and the search result can then be filtered to select only information on legislation. The information is free to companies and organizations in developing countries, but they have to register first.

http://www.cbi.nl/marketinfo/cbi/?

Product liability

The directive on Product Liability of 1985 (amended in 1999)[35] covers any product as well as raw materials from EU origin or imported. It makes the producer liable for damage caused by a defect in the product when the injured party proves the damage, the defect and the causal relationship between the two. This applies even if the producer proves he/she was not negligent. "Producer" means the manufacturer of a finished product, the producer of any raw material or the manufacturer of a component part and any person who, by putting his/her name, trademark or other distinguishing feature on the product,

---

[33] Source: www.eco-labels.org.

[34] http://www.publix.com/about/qa/QAProducts.do

[35] Directive 85/374/EEC, amended by 1999/34/EC, see http://europa.eu.int/eur-lex/lex/LexUriServ/site/en/consleg/1985/L/01985L0374-19990604-en.pdf for consolidated text.

presents him/herself as its producer. All importers are considered to be producers within the meaning of the directive.[36]

General Food Law
Regulation EC 178/2002, the General Food Law, was adopted in 2002. It established the European Food Safety Authority (EFSA: http://www.efsa.eu.int) and laid down procedures for food safety issues which included the reorganization of the regulatory system.

Parts of the law became effective immediately, while other parts were implemented at later dates. For the official text, see:

http://europa.eu.int/eurlex/pri/en/oj/dat/2002/l_031/l_03120020201en00010024.pdf

Food hygiene
Since January 2006 the new regulation 852/2004 on the hygiene of foodstuffs has applied. The main difference with the old Directive 93/43/EEC on Food Hygiene[37] is that the new regulation also applies to foreign operators who produce food for consumption in the European Union.

According to the new regulation, primary producers of foodstuffs must comply with the hygiene provisions in part A of Annex I. For producing plant products, this includes: keeping equipment, facilities and transport means clean; using clean water whenever necessary to prevent contamination; and ensuring that staff are in good health and undergo training on health risks. Food business operators other than primary producers must also comply with the hygiene requirements in Annex II, which are more detailed and prescriptive. (For example they have to have an adequate number of flush lavatories.)

The second group must also implement procedures based on the HACCP principles and provide the competent authority with evidence of this. Registration of both groups will have to take place with the appropriate competent authority.

http://europa.eu.int/eur-lex/pri/en/oj/dat/2004/l_226/l_22620040625en00030021.pdf

Pesticide residues
The European Union is in the process of harmonizing all of its Maximum Residue Limits (MRLs) and is expected to reach complete harmonization soon. Harmonized MRLs for fruits and vegetables are published in the consolidated text of EC Directive 76/895/EEC. The latest update of this consolidated text is from 5 January 2004:

http://europa.eu.int/eur-lex/en/consleg/pdf/1976/en_1976L0895_do_001.pdf

The European Union also provides lists of harmonized MRLs by pesticide, by crop group and by commodity. The lists were last updated in November 2004:

http://europa.eu.int/comm/food/plant/protection/pesticides/index_en.htm

Another source of MRLs is the Pesticides Action Network (PAN) Pesticides database. A country search gives pesticide registrations and Prior Informed Consent information per country: http://www.pesticideinfo.org/Search_Countries.jsp#Europe%20and%20 Central%20Asia

Genetically modified organisms
Although commercial production of genetically modified fruits and vegetables is still very limited (e.g. only tomato and pepper in China[38]), many field trials are being conducted around the world on a variety of vegetables and several fruits. Therefore, in the future, EU legislation on GMOs and labelling of food derived from GMOs may also become relevant for fruit and vegetable producers.

---

[36] http://www.newapproach.org/
[37] Food hygiene directive consolidated text:
http://europa.eu.int/eur-lex/en/consleg/pdf/1993/en_1993L0043_do_001.pdf
[38] Runge, C.F. 2005 Genetically Modified Crops/foods: Diffusion and Development in 2005, presentation at FAO, Rome.

Regulation 1829/2003 regulates the placing on the market of food and feed products containing or consisting of GMOs and also provides for the labelling of such products for the final consumer. In practice, it is the seed companies developing GMO varieties that invest in the EU approval process.

http://europa.eu.int/eur-lex/pri/en/oj/dat/2003/l_268/l_26820031018en00010023.pdf

Regulation 1830/2003 introduces a harmonized EU system to trace and label GMOs and to trace food and feed products produced from GMOs.

http://europa.eu.int/eur-lex/pri/en/oj/dat/2003/l_268/l_26820031018en00240028.pdf

Irradiation

Irradiation is regulated by the EU Directives 1999/2/EC (ionizing radiation standards) and 1999/3/EC (list of foodstuffs that may be irradiated). This list is very short: only dried aromatic herbs, spices and vegetable seasonings are allowed to be irradiated. National legislation may allow other product groups to be irradiated. The words "irradiated" or "treated with ionizing radiation" must appear on the label or packaging.

http://europa.eu.int/eur-lex/pri/en/oj/dat/1999/l_066/l_06619990313en00160022.pdf

http://europa.eu.int/eur-lex/pri/en/oj/dat/1999/l_066/l_06619990313en00240025.pdf

4.3. CORPORATE STANDARDS

Global Food Safety Initiative (GFSI)

GFSI is a food retailer initiative and was launched in May 2000. It is based on the principle that food safety is a non-competitive issue, as any potential problem arising may cause repercussions in the whole sector. The Initiative is facilitated by CIES-The Food Business Forum. In July 2005, the GFSI-Foundation was established as a separate legal entity.

The mission of GFSI is to strengthen consumer confidence in the food they buy in retail outlets. GFSI did not produce a new standard but a benchmark model. The model identifies three key elements that a food safety standard should contain:

- a food safety management system
- good agricultural practices and/or good manufacturing practices and/or good distribution practices (depending on the types of entities to be certified under the benchmarked standard)
- an HACCP system, based on or equivalent to the Codex Alimentarius HACCP.

The benchmark model is contained in the GFSI Guidance Document, of which the fourth edition was published in July 2004. Currently there are four standards that comply with the GFSI guidelines:

1. The BRC Global Standard – Food
2. International Food Standard
3. Dutch HACCP code
4. SQF 2000 code

In theory, retailers should accept certification against one of these standards, which is considered equivalent to certification against any of the remaining ones. However, in practice, retailers prefer certification against their "own" standard. For example, suppliers to European retailers may need BRC certification for the UK market, IFS certification for the German and French markets and Dutch HACCP certification for the Dutch market.

The GFSI has also published a document Implementing Traceability in the Food Supply Chain in response to the EU traceability requirements. GFSI is also undertaking a project on the coordination of Good Retail Practices.

For more information, please see: www.ciesnet.com,

http://www.ciesnet.com/pdf/globalfood/GFSI_Year_Book_2004_FINAL.pdf

http://www.ciesnet.com/pdf/globalfood/2005-09-newsletter.pdf

British Retail Consortium (BRC)

BRC is the trade association that represents 90 percent of the retail sector in the United Kingdom. In the United Kingdom, under the terms of the Food Safety Act 1990, retailers have an obligation to take all reasonable precautions and exercise all due diligence to avoid failure in the development, manufacture, distribution, advertising or sale of food products to the consumer. This obligation led to the introduction of various standards. (See also http://www.brcglobalstandards.com).

BRC Global Standard - Food

Formerly called The BRC Food Technical Standard, the latest version was published in January 2005 as the BRC Global Standard – Food, Issue 4, and came into effect in July 2005. The standard covers topics such as: the HACCP system; quality management; factory environment standards; and product and process control. Further details on the standard's requirements are not available since the standard needs to be purchased.

Scope: The standard is meant for use by food service companies, catering companies and food manufacturers. It covers the supply of retailer-branded products, branded food products and processed or prepared food or ingredients. The standard is intended to be implemented by suppliers of retailers recognizing the standard (i.e. BRC members).

Certification system: Suppliers need to be certified by an EN45011/ISO65-accredited certification body. The standard provides certification to be awarded at two levels: Foundation Level and Higher Level.

Labelling: Certified suppliers cannot use the BRC Logo unless they sign up (and pay) to use the Global Standards Directory. In this case, they may use the logo in company brochures, etc. but not on any commercial product.

BRC-IoP Global Standard - Packaging

Manufacturers have an obligation to put appropriate systems and controls in place to ensure the suitability of their packaging for safe food use. This Standard is most applicable to manufacturers of food contact packaging. The standard came into effect in March 2005.

The standard covers the following topics: organization; hazard and risk management system; technical management system; factory standards; contamination control; personnel; risk category determination; and the evaluation protocol.

BRC/FDF Technical Standard for the Supply of Identity-preserved Non-Genetically Modified Food Ingredients and Products

The Food and Drink Federation (FDF) is the voice of the UK food and drink manufacturing industry. Together with the BRC they developed this standard for sourcing on non-GMO soybeans and maize. In a later stage it may also be applied to other products as it "provides a framework for identity-preserved systems for other commodities."

The International Food Standard (IFS)

In 2002, German food retailers from the HDE (Hauptverband des Deutschen Einzelhandels) developed a common audit standard called International Food Standard (IFS). In 2003, French food retailers and wholesalers from the FCD (Fédération des entreprises du Commerce et de la Distribution) joined the IFS Working Group and have contributed to the development of IFS version 4.

IFS has been designed as a uniform tool to ensure food safety and to monitor the quality level of producers of retailer-branded food products. The standard can apply in all steps of the processing of foods subsequent to their agricultural production. The HDE Trade Services GmbH was assigned to take over the administration of the IFS. IFS version 4 is available in German, English and French at a price of €39.00, and in

Polish, Dutch, Italian and Spanish for €59.00 (plus shipping and handling).

The content of the standard can be summarized as follows:

- Management of the Quality System: the HACCP system, team and analysis; the quality manual; and the obligation to keep reports and documents.
- Management responsibility: management responsibility, commitment and review; and customer focus.
- Resource management: personnel issues (hygiene, medical screening); and staff facilities.
- Product realization: adherence to specifications for products, factory environment, pest control, maintenance and traceability; GMO and allergen.
- Measurements, analyses, improvements: e.g. internal audit, all types of controls during production steps; product analysis; and corrective actions.

There are criteria at three levels: Foundation level (minimum requirements); Higher level; and Recommendations (for demonstration of "best practice"). The IFS compendium of doctrines, which summarizes clarifications about some IFS requirements, applies from 1 June 2005.

Certification may only be done by certification bodies that are accredited against EN 45011 (ISO 65) on IFS. Only IFS-authorized auditors who have passed a written and oral examination and have competence in the specific sector can audit against the standard. A BRC audit cannot be used for IFS certification.

An Internet audit portal has been developed. Every passed audit report with an IFS certificate will be included in the online database. Only the name and the address of the audited company/branch are published. The audited company can choose to make more details of the audit available to customers by giving them specific access.

Market penetration: German and Swiss retailers in the HDE-Committee for Food Law and Quality Assurance and French retailers in the FCD-Quality Committee " support" the standard.[39] It is not clear if "support" also means that certification against the standard is required from all suppliers of their retailer-branded products. The application of IFS in Poland, Austria, Belgium, Netherlands, United Kingdom and Italy is being discussed with wholesalers and retailers and their federations. http://www.food-care.info/

The Dutch HACCP code

The Requirements for an HACCP-based Food Safety System or the Dutch HACCP code was developed by a committee of experts represented by retailers, consumers, governmental food safety inspection agencies and producer associations. A food business operator may use the code to develop its HACCP-based Food Safety System.

In addition to implementing an HACCP system, the Dutch HACCP code requires the implementation of relevant elements of ISO 9000 to support the food safety management system and some GAP and GMP elements. The requirements reference the Codex guidelines on HACCP, the EU Directive on Food Hygiene and specific food safety legislation. Where specific requirements do not exist, the Codex General Principles on Food Hygiene are applied.

Certification bodies wishing to certify operators against the code must be accredited against ISO/IEC Guide 62:1996 (EN 45012) and the Standards for auditing (ISO 10011, Parts 1, 2 and 3). The Dutch HACCP code also provides rules for the certification process (e.g. minimum auditor time). The standard and the requirements for certification bodies (in Dutch) can be downloaded for free from: http://www.foodsafetymanagement.info.

---

[39] HDE: Metro AG, REWE, EDEKA, Aldi, Tengelmann, AVA, Tegut, Globus, Markant, Lidl, Spar, COOP (Switzerland) and Migros (Switzerland). FCD: Auchan, Carrefour, EMC - Groupe Casino, Metro, Monoprix, Picard Surgelés, Provera (Cora and Supermarchés Match).

Market penetration: The Dutch HACCP code is the smallest of the three European-based systems benchmarked by the CIES Global Food Safety Initiative. It is mainly supported by the Dutch retailers.

The Safe Quality Food Program (SQF 1000 and SQF 2000)

The SQF Program originated in Australia but is now administered by the SQF Institute, a division of the Food Marketing Institute in the United States. The Food Marketing Institute has 1 500 food retail and wholesale member companies, operating in approximately three-quarters of all food retail store sales in the United States.

The SQF Program is based on the principles of HACCP, Codex, ISO and Quality Management Systems. The SQF 1000 Code is designed specifically for primary producers. It combines Good Agricultural Practices with Food Safety and Food Quality Plans. The SQF 2000 Code is for implementation by food manufacturing and distribution sectors. It combines Good Manufacturing Practices with the Food Safety and Food Quality Plans.

A number of modules are provided as voluntary options to suppliers whose markets require additional assurances. They include Social Accountability, Environment, Animal Welfare, Organic and Bioterrorism (no content of these modules was found on the web).

Market penetration: Most of the approximately 1 600 SQF1000-certified companies are Australian, with 12 from the Republic of Korea and 1 (Dole Stanfilco) from the Philippines. Recently more and more US companies have become certified (about 250).

SQF2000 certification is much more international. There are many certifications issued in the Netherlands, Thailand, Korea, Saudi Arabia, India and China. Furthermore, one or several companies are certified from Japan, Viet Nam, United Arab Emirates, Singapore, Sri Lanka, Philippines, Peru, New Zealand and Mauritius.

Source: http://sqfi.com/

Scientific Certification Services (SCS) – Certiclean and NutriClean certification

SCS is an environmental certification company that began offering certifying services in 1984. SCS offers several certification programmes, including CertiClean. SCS assists with the development and documentation of HACCP-based food safety management programmes. Implementing and executing the food safety programme successfully for 60 days will qualify a company's programme for certification through the CertiClean programme. Certified companies are subject to quarterly audits to maintain certification. Companies outside the United States have also been certified.

SCS offers also a Vendor Verification Program designed for distributors, processors, packers, and retailers who need to assess the food safety practices of their suppliers. Vendor verification is essentially a Good Manufacturing Practices check on a client's suppliers.

SCS offers also two pesticide residue-testing programmes. The Dock programme tests samples of incoming produce, both domestically grown and imported, for grocery retailers, at their distribution dock, to ensure that pesticide levels are within US governmental tolerances. In addition, SCS has developed the NutriClean - Pesticide Residue Free Certification programme. For this scheme, maximum residue limits are all set to the limit of detection as defined by the US Government. Certified products may carry the NutriClean label.

Market penetration: Although most CertiClean and NutriClean certifications are of US operators, some Central American operators have been certified as well (e.g. Guatemala and Mexico). No information on buyer recognition of SCS programmes could be found. With respect to consumer response, Lamb et al. claim that focus-group research found that consumers actually reacted negatively to any mention of pesticides, and pesticide-free labelling was therefore largely abandoned in the mid-1990s.40 http://www.scscertified.com/foodAgriculture/

European Food Safety Inspection Service (EFSIS) food standards

EFSIS is a UK-based certification body with its core business in the livestock sector. It certifies worldwide against many of the standards discussed in this publication. EFSIS has also developed two of its own standards for packers and manufacturers: The Standard for companies supplying food products and The Health & Safety Standard. No indication of their content and certification criteria or level of recognition from buyers could be found on the EFSIS web site http://www.efsis.com.

One of the EFSIS standards was benchmarked by the GFSI and found to be in compliance with its guidance document. However, after the last revision of the GFSI guidance document, EFSIS withdrew its standard from the benchmark process.

American Institute of Baking (AIB) standards and certification programmes

AIB is a non-profit corporation founded in 1919 by the North American wholesale and retail baking industries. The Institute currently serves many segments of the food processing, distribution, food service and retail industries worldwide.

AIB has developed many standards, but also offers certification against standards from other organizations, such as the BRC standard, HACCP and organic certification. AIB also offers help in designing nutrition labels according to US law. Based on basic ingredient information, AIB calculates nutritional value per 100 grams and per serving.

AIB's own standards on food safety include:
- Consolidated Standards for Agricultural Crops
- Consolidated Standards for Food Contact Packaging Manufacturing Facilities
- Consolidated Standards for Food Distribution Centres
- Consolidated Standards for Food Safety
- Consolidated Standards for Fresh-cut Produce
- Consolidated Standards for Fresh Produce & Fruit Packinghouses
- Consolidated Standards for Nonfood Contact Packaging Manufacturing Facilities

These standards (also available in Spanish) form the basis of an AIB International food safety/hygiene audit, whose purpose is internal learning by the audited company. This includes an optional rating system that provides management with an index of how well a facility is complying with food safety regulations as well as with the established internal standards set by the individual company. It is also possible for a company to write its own standard and have an audit of its factories conducted against this standard. AIB International conducts audits worldwide.

Furthermore, AIB International offers the Gold Standard Certification Program, an integrated quality system for the food industry. It consists of a Good Manufacturing Practices Audit Qualification; HACCP Validation and Verification; and a Quality Systems Evaluation. Certified facilities may use the AIB gold standard logo. http://www.aibinternational.com/

Certification of implementation of FDA-guidelines

Many US retailers require third-party verification of implementation of the FDA Guide to Minimize Microbial Food Safety Hazards for Fresh Fruits and Vegetables of 1998. However, this requirement is often only directly communicated to the suppliers in question and not made public. Therefore, it is not possible to provide a list of US retailers requiring FDA-

---

40 Lamb, J.E., Velez J.A. and Barclay R.W. 2004. The Challenge of Compliance with SPS and Other Standards Associated with the Export of Shrimp and Selected Fresh Produce Items to the United States Market. World Bank Agriculture and Rural development Discussion paper. (page 50)

Guide certification. Various certification programmes to satisfy this demand have been developed by certification bodies.

ProSafe Certified Program by Davis Fresh

An example of a certification programme to verify the implementation of the FDA Guide is ProSafe. Davis Fresh is an independent consultancy company in the fresh produce industry. The company offers the ProSafe Certified Program, a three-step programme that includes: Food Safety Risk Assessment, Food Safety Risk Mitigation and Final Audit and Certification. http://www.davisfreshtech.com/prosafe/index.html

From the ProSafe master listing of buyers, it can be presumed that certification is "recognized" by: Albertsons, DFT, Kroger, MARKON, Raley's Supermarkets, Royal Ahold, Safeway, Sainsbury, Sam's Club, SYSCO, Times Supermarkets, Waitrose and Wal-Mart. This does not mean that these retailers have made certification obligatory.

GAP Certification Program of PrimusLabs

Another example is the GAP Certification Program by PrimusLabs: http://www.primuslabs.com.

PrimusLabs offers a programme with three phases. In the first phase, suppliers must develop operation-specific manuals according to PrimusLabs procedures and train staff. For growers and/or packers of fruits and vegetables, these can be:

- GAP Manual to comply with USA (i.e. the FDA guide)
- GreenHouse Manual

Manual of Packing Operations and Food Safety Issues (including GMP, HAZARD PLAN and Sanitation Standard Operating Procedures).

In the second phase, they must conduct self-audits. In the third phase, the supplier must contract an approved auditing firm to verify implementation of the FDA Guide 1998. http://intranet.primuslabs.com/igap/

With some variations, this programme is recognized, promoted or required by Avendra, Publix, Rubios, Albertsons, Subway and Safeway. Avendra is a large procurement company for the hospitality industry in North America and the Caribbean, and began the programme in 2001. Publix Super Markets is the largest and fastest-growing employee-owned supermarket chain in the United States. Publix began implementation of the food safety programme in 2002. Rubios Fresh Mexican Grill has over 150 restaurants and started the programme in 2004. In addition to implementation of the FDA guide, Avendra requires HACCP implementation for production, storage and distribution and Albertsons encourages HACCP implementation from processors.

Albertsons claims to be the second-largest grocery chain in the United States. Albertsons does not formally require the programme but recognizes levels of commitment according to the progress in implementation. It is not clear what consequences or benefits each recognition level brings for the supplier. Apart from PrimusLabs, Albertsons also directs its suppliers to other sources of assistance, such as the FMI SQF programme, the International Fresh-cut Produce Association (IFPA), the Produce Marketing Association (PMA) and the United Fresh Fruit and Vegetable Association (UFFVA)

Subway is the world's largest sandwich chain, with more than 21 000 restaurants in 75 countries. In 2001 and 2002 Subway worked with PrimusLabs to implement its Gold Standard Program for its fresh produce suppliers. In 2001 it asked its suppliers (produce buyers) to register all of their food safety practices utilizing the PrimusLabs Document Development Program. These suppliers, including farmers, also needed to carry out self-audits. In 2002 third-party audits were required for all produce sold under the Subway label. The requirements, against which audits take place, were explained in manuals developed by PrimusLabs. There is a manual for every step in the supply chain: farm; cooling/cold store; packing house; and trace recall for the shipper.

Safeway

Since 1999 Safeway has required third-party verification of compliance with GAP and GMP for growers and shippers of produce that had been implicated in food-borne illness outbreaks. Audit parameters included: ranch history; adjacent land use; harvest practices; transportation; and food production and handling during growing, harvesting, coding and storage. Safeway had already identified some Independent third-party auditors, but would accept other auditors as well. Suppliers were free to communicate audit results to other customers.[41]

Since May 2004 all new produce suppliers are being asked to sign an extra agreement on top of the normal supplier-Safeway agreement. One requirement of the extra agreement is third-party verified compliance with GAP and GMP as described in the FDA Guide to Minimize Microbial Food Safety Hazards for Fresh Fruits and Vegetables and in Title 21 of the Code of Federal Regulations (see: www.safeway.com).

In June 2004, Safeway sent a letter to its suppliers of refrigerated and frozen products, in which it communicated mandatory temperature receiving requirements, including for bagged and pre-packaged salads and fresh herbs.

Merchants Distributors

Merchants Distributors, Inc. (MDI) is a privately owned wholesale grocery store distributor in the United States supplying over 600 retail food stores. MDI is also implementing a food safety programme, but its suppliers are free to choose a consultancy or request assistance from a specific programme of a certification body. No deadlines for third-party audits are set.

---

[41] Source: http://intranet.primuslabs.com/igap/letters/safeway.htm. Primuslabs was one of identified ITPA.

5. Sustainable agriculture and good agricultural practices (GAP)

This chapter gives an overview of standards for sustainable agriculture or good agricultural practices (GAP). The standards are "mixed", in the sense that they combine environmental, social and/or food safety standards. ("Purely" environmental standards are discussed in the next chapter. Organic agriculture is discussed in chapter 7. Standards that focus mainly on labour conditions and/or other social issues are discussed in chapter 8.)

5.1. INTERNATIONAL POLICY INSTRUMENTS

The Earth Summit and Agenda 21

The Earth Summit

In 1983, the United Nations appointed an international commission to propose strategies for "sustainable development". Its report, Our Common Future (also known as the The Brundtland Report), was published in 1987 and had worldwide impact on policy development. At the 1992 United Nations Conference on Environment and Development (UNCED, or the Earth Summit) in Rio de Janeiro, the international community adopted Agenda 21, a global plan of action for sustainable development. The Commission on Sustainable Development (CSD) was created in December 1992 to ensure effective follow-up of UNCED. It falls under the United Nations Department for Economic and Social Affairs (UN-DESA): http://www.un.org/esa/sustdev/csd/csd.htm.

Agenda 21 designated nine sectors of society, known as Major Groups, as critical for the development and implementation of policies for sustainable development.[42]

Agenda 21, Chapter 4 and the Marrakech Process

Chapter 4 of Agenda 21 is titled Changing consumption patterns and states that governments should encourage the expansion of environmental labelling and other environmentally related product information programmes designed to assist consumers to make informed choices.

The Marrakech Process is a series of international and regional meetings organized by UN-DESA and UNEP to develop a 10-Year Framework of Programmes for Sustainable Consumption and Production. It is named after the first international expert meeting organized by the UN Social and Economic Council in Marrakech in 2003. One of the discussion topics of the meeting was ecolabelling (mainly national ecolabelling schemes for industrial products, but also fair-trade labelling), and the meeting requested UNEP to explore the feasibility of a partnership with the Global Eco-labelling Network.

http://www.un.org/esa/sustdev/sdissues/consumption/MarrakechReport.pdf

After a series of regional meetings, a second international meeting under the Marrakech Process on the 10-Year Framework was held in September 2005 in Costa Rica. The Working Group on Sustainable Consumption and Product Development considered three key issues: institutional procurement; sustainable product design; and sustainable lifestyles. Institutional procurement officers would benefit from a product database, including information on products of small- and medium-sized enterprises

---

[42] The nine Major Groups are: Farmers, NGOs, Indigenous Peoples, Workers and Trade Unions, Business & Industry, Local Authorities, the Research and Technological Community, and Women and Youth.

and on fair-trade products. Existing databases of UNEP and IGPN (International Green Purchasing Network) could be enhanced.

The Working Group on Regional and National Strategies for Sustainable Consumption and Production agreed on the need to link Sustainable Consumption and Production to the Millennium Development Goals. This would highlight and demonstrate the contribution of Sustainable Consumption and Production to reducing poverty (e.g. fair-trade and organic agriculture can increase income and quality of life of farmers in developing countries).

> http://www.un.org/esa/sustdev/sdissues/consumption/Marrakech/conprod10Y2ndafr.htm
> UNEP/IAPSO product criteria database (for sustainable procurement):
> http://www.uneptie.org/pc/sustain/policies/green_find.asp
> IGPN: http://www.gpn.jp/igpn/index.html

Agenda 21, Chapter 14: Sustainable Agriculture and Rural Development (SARD)
Chapter 14 of Agenda 21 deals specifically with Sustainable Agriculture and Rural Development (SARD). FAO was appointed task manager to monitor the implementation of Chapter 14. Following an electronic conference, a multistakeholder forum on SARD was organized concurrently with the FAO Committee on Agriculture (COAG) in April 2001. The dialogue was continued at the World Food Summit Five Years Later (Fyl) in June 2002 and progress was reported at the World Summit on Sustainable Development (WSSD) in Johannesburg, 2002.

The FAO SARD-Interdepartmental Task Force has identified three programme thrusts:
1. Sustainable Livelihoods
2. Sustainable Intensification of Integrated Production Systems and
3. Integrated Natural Resources Management.

For Thrust 2, COAG endorsed that the development and mainstreaming of FAO's multisectoral Good Agricultural Practices (GAP) approach and other action to enhance sustainability and safety of food chains should be a priority.

Good Agricultural Practices (GAP)
Adoption of agricultural practices which protect the environment and ensure the quality and safety of food as well as increasing productivity should enable farmers to increase their incomes from existing markets and take advantage of new market opportunities, thus achieving sustainable improvements in their livelihoods.

FAO's approach to Good Agricultural Practices (GAP) is non-prescriptive, and would not lead to the development of new international mandatory standards or codes. Further, it does not affect the definition or scope of GAP as they appear in existing texts. Instead, local-level GAP defined by concerned stakeholders would draw inspiration both from existing international texts and from the broader GAP principles. These broad principles were presented to the Committee on Agriculture in 2003 and relate to: soil; water; crop and fodder production; crop protection; animal production; animal health and welfare; harvest and on-farm processing and storage; energy and waste management; human welfare, health and safety; and wildlife and landscape (see Annex 3 of http://www.fao.org/DOCREP/MEETING/006/Y8704e. HTM#P53_14805).

GAP work in FAO has already proved effective in helping production experts to better understand how to incorporate food safety and quality requirements into their technical recommendations, and in helping food safety experts to better understand broader sustainability considerations at farm level.

5.2.   NATIONAL REGULATIONS

United States

As described in Chapter 4 on food safety, the FDA Guide to Minimize Microbial Food Safety Hazards for Fresh Fruits and Vegetables of 1998 contains guidance on good agricultural practices. However, because these practices focus entirely on the reduction of microbiological contamination, they are considered food safety guidelines for the purpose of this publication.

European Union

There is no legislation on good agricultural practices in the European Union that has any effect on foreign suppliers, except for specific food safety regulations discussed in chapter 4.

5.3.   CORPORATE STANDARDS

EurepGAP

Introduction

EurepGAP is driven by European retail chains, more specifically the members of the Euro-Retailer Produce Association (EUREP). FoodPLUS GmbH, a commercial company, serves as legal owner of the normative documents of the EUREP Good Agricultural Practices (EurepGAP) standards, and hosts the EUREP Secretariat.

The EurepGAP Fruits and Vegetables standard was developed by the EurepGAP Technical Committee – Fruits and Vegetables. This Committee used to be dominated by retailers, but now consists equally of retailers and supplier representatives. In this text, the term EurepGAP is used to indicate "EurepGAP Fruits and Vegetables".

EurepGAP has the declared aim of increasing consumer confidence in the safety of food and of harmonizing buyers' requirements for food hygiene and for Maximum Residue Limits for pesticides. The norms also address some environmental and labour issues although in practice only a minority of the control points relate to these issues.

Requirements

The EurepGAP® Standard has a checklist of 49 major control points, 99 minor control points and 62 recommendations. Producers must comply with all applicable major control points and with 95 percent of the applicable minor control points. Producers must also provide justification for considering a control point not applicable, and some control points cannot be excluded.

Requirements applicable to all stages in the production process are those on traceability, record keeping and self-inspection, waste and pollution management and worker health and safety. Other requirements address specific production phases: varieties and rootstocks; site history and management for site selection; soil and substrate management; fertilizer use; irrigation and crop protection for crop production; harvesting; and produce handling.

Certification bodies wishing to certify against EurepGAP have to be accredited against ISO65/EN45011 and FoodPLUS. A particularity of the EurepGAP system is the possibility of issuing non-accredited certificates. Each applicant certification body is given a period of six months to complete the required accreditation, which includes the issuing of non-accredited certificates as a practical exercise. In communications with the retailer, such non-accredited certificates are said to also be accepted. In addition to the certification costs, certified producers have to pay a small annual fee to FoodPLUS (around €25 per year).

A grower can have EurepGAP-registered and -non-registered products. However, each registered crop or product must be completely covered by a EurepGAP certificate – e.g. if a farm grows cauliflower, and part of it is marketed through other channels, all of the cauliflower must still be produced and certified according to EurepGAP.

Farmer associations that have already implemented an existing farm assurance scheme with third-party verification can benchmark that scheme against EurepGAP. If the farm assurance scheme is accepted as equivalent and is accredited, the farm audit for that scheme would serve as a EurepGAP audit as well. Requests for a benchmarking procedure should be made by the legal owner of the scheme. In the case of government ownership, this means the request should come from a government representative.

It is also possible for a so-called Produce Marketing Organization (PMO) to obtain a group certification. A PMO can be a cooperative or other group of growers that has a legal entity. This entity would take over responsibilities, through an internal control system, of EurepGAP implementation for the associated and contracted growers. Detected non-compliance of one farmer in the group may lead to de-certification of the entire group.

Penetration and market

There is no product label associated with EurepGAP certification and no premium. EurepGAP is currently preparing the rules and prerequisites for carrying a reference to EurepGAP at individual box level. The market for products from EurepGAP-certified produce consists of the retailers promoting the certification. Of Europe's ten largest chains, three promote EurepGAP (Tesco, Metro-Group and Ahold). Certification will not be a guarantee for being listed by those supermarkets, but may become a prerequisite. It was said that some retailers would require EurepGAP certification for fresh fruits and vegetables as early as January 2003. According to a guidance document by the Australian Government, Albert Heijn in the Netherlands requested their suppliers to implement EurepGAP® by 1 January 2003. In the United Kingdom market, Sainsbury's instructed its suppliers to commence implementation of EurepGAP® by 1 January 2004.[43]

However, in spring 2003, information on the EurepGAP web site read: "Some retailers are saying that all their suppliers must be EurepGAP-certified by 2004. Others do not have a deadline, but will in time question why preferred suppliers are not EurepGAP-certified and perhaps review their decision to do business with them."

Since 2003 no statements about implementation deadlines for suppliers have been made public (at least the author of this report could not find any). Thus, there is considerable uncertainty about the degree of flexibility in implementation schedules and about if, when and which retailers really will exclude suppliers form their preferred lists if they have not yet been certified.

Lamb et al. state that EurepGAP is also increasingly accepted in the United States in lieu of other third-party certification such as SQF1000.[44]

The number of certifications in developing countries has increased sharply during the last two years, with certificates issued in 37 different developing countries. There are now also a number of certification bodies with headquarters in developing countries (see Table 1). Of course, European-based certifiers also certify in developing countries.

There are only three farm assurance schemes that completed the (re-)benchmarking procedure and were fully recognized. They are from Austria, Spain and Chile. Re-benchmarking against the new version of 2004 is about to be completed for another four schemes in the United Kingdom, Spain and New Zealand, and another five European schemes are at various stages of the benchmarking process.

KenyaGap and the Mexico Supreme Quality programme have also requested to start benchmarking procedures. The Fresh Produce Exporters Association of Kenya

[43] Commonwealth of Australia. 2004. Guidelines for implementing EUREPGAP for Australian Fruits and vegetable Producers. Available at: http://www.daff.gov.au/corporate_docs/publications/pdf/food/eurepgap_guidelines.pdf.
[44] Lamb, J.E., Velez J.A. and Barclay R.W. 2004. The Challenge of Compliance with SPS and Other Standards Associated with the Export of Shrimp and Selected Fresh Produce Items to the United States Market. World Bank Agriculture and Rural Development Discussion paper. (page 39 and 50)

(FPEAK) established a National Technical Working Group to develop smallholder

Table 1 - EurepGAP-accredited certification bodies in developing countries

EurepGAP Interpretation Guidelines and sustainable compliance modalities for small-scale farmers that will be used in Kenya. The results will be incorporated into the revision of KenyaGap, which will then be benchmarked against EurepGAP.

The Mexico Supreme Quality logo has been implemented to drive a promotional programme in key export markets. The standards implemented to use the logo have laid the foundations for EurepGAP compliance.

In a further effort to promote EurepGAP implementation outside Europe, in May 2005 EurepGAP chairman Nigel Garbutt visited China, where he met vice-chairperson Cheng Fang of the Certification and Accreditation Administration of the People's Republic of China (CNCA).

The Farm Accreditation Scheme of Malaysia is developing MalaysiaGap in order to become benchmarked and recognized as equivalent by EurepGAP.

Complaints about EurepGAP in SPS Committee

Private-sector standards have never been discussed in the SPS Committee, although they have been raised in the Technical Barriers to Trade (TBT) Committee. Saint Vincent and the Grenadines, supported by Jamaica, Peru, Ecuador and Argentina, complained in the SPS committee meeting of June 2005 that EurepGAP's requirements were tougher than the EU requirements. The European Union replied that it was not in a position to intervene because the business-sector organizations claim they are reflecting consumer demand.

Some called for SPS Article 13 to be clarified. This article states that:

"(..) Members shall take such reasonable measures as may be available to them to ensure that non-governmental entities within their territories, as well as regional bodies in which relevant entities within their territories are members, comply with the relevant provisions of this Agreement. (..)" (A similar requirement in the TBT Agreement was discussed in chapter 1.1).[45]

Sustainable Agriculture Initiative (SAI) Platform
Introduction
The SAI-Platform was founded in 2002 by three major global food industry companies – Unilever, Nestlé and Danone – to actively support the development of sustainable agriculture and to communicate it worldwide. By 2005, 19 food industry companies were already members of the SAI-Platform[46]. The platform defines sustainable agriculture as a productive, competitive and efficient way to produce agricultural raw materials, while at the same time protecting and improving the natural environment and socio-economic conditions of local communities.

The SAI-Platform aims for recognition and implementation of sustainable practices for mainstream agriculture (not niche markets) on a worldwide scale. The individual SAI-Platform members are free to decide whether to participate in assessment processes, and are free to take any specific action, such as with respect to implementation. The SAI-Platform held its first General Assembly in April 2003.

There are Working Groups on Cereals, Dairy, Palm Oil, Fruits, and Potatoes and Vegetables. Draft guidelines are tested through pilot projects and discussed with stakeholders. Furthermore, the Steering Group on the social and economic benefits of

---

[45] WTO 2005 news item, 4 July 2005.

[46] They were: Unilever, Danone Group and Nestlé. Furthermore Campina, CIO (Consorzio Interregionale Ortifrutticoli), Danisco, Dole, Ecom, Efico, Findus, Fonterra, Friesland Foods, Kraft, McCain, McDonald's, Neumann Kaffee Gruppe, Sara Lee/DE, Tchibo, Volcafé.

Africa		
Africert LTD	Kenya	December 2004
Center of Organic Agriculture in Egypt (COAE)	Egypt	August 2004
Egyptian Center of Organic Agriculture	Egypt	April 2004
Perishable Products Exports Control Board (PPECB)	South Africa	April 2002
SABS Certification (Pty) Ltd	South Africa	August 2002
SGS ICS (Pty) Ltd	South Africa	May 2005
Latin America		
BSi Inspectorate de Argentina S.A.	Argentina	June 2004
BVQI do Brasil Sociedade Certificadora Ltda	Brazil	March 2004
Eco-LOGICA	Costa Rica	June 2005
IBD-Institut Predefined report from http://europa. eu.int/comm/agriculture/ofis_public/pdf/r8_0000_en.pdf, generated August 2005 o Biodinamico	Brazil	April 2005
IRAM-Instituto Argentino de Normalización y Certificación	Argentina	May 2003
LATU Sistemas S.A.	Uruguay	December 2001
Organización Internacional Agropecuaria	Argentina	April 2004
Serviço Brasileiro de Certificaçoes Ltda	Brazil	May 2005

Table 1 - EurepGAP - accredited certification bodies in developing countries

sustainable agriculture aims to compile concrete information on the advantages and disadvantages (costs/benefits, risks, etc.) of conversion to sustainable agriculture.

Before starting the SAI-Platform, Unilever was already working in pilot projects on good agriculture practices and on sustainability indicators. The ten indicators are: soil fertility and health; soil loss; nutrients; pest management; biodiversity; product value; energy; water; social and human capital; and local economy. Unilever guidelines for best practices have been developed for tea (plantations and smallholders), palm oil, peas and spinach, and are being developed for tomato, rapeseed and sunflower. The guidelines are specific for the country in which the pilot project that developed them operates or operated.[47]

Requirements

The SAI-Platform's Working Group on Potatoes and Vegetables consists of McCain (chair) and Consorzio Interregionale Ortofrutticoli (CIO), Findus, Kraft, McDonald's, Nestlé and Unilever. They have just published (July 2005) draft principles containing chapters on:

- Sustainable farming system (site selection; planting material; IPM; sustainability management system; access to information and support services)
- Economic sustainability (safety with MRL targets below legal MRLs, quality and transparency; financial structure; relation to market; diversification)
- Social sustainability (labour conditions; training; strengthening local economy)
- Environmental sustainability (soil conservation; water conservation; biodiversity conservation; integrated waste and crop by-product management; energy conservation; air conservation).

Of note is that HACCP is not mentioned. The principles in chapter 3 are based on ILO Conventions 111 (on Discrimination), 100 (on Equal Remuneration), 29 and 105 (on Forced Labour), 87 (on Freedom of Association) and 98 (on the Right to Organize and Collective Bargaining), 138, 182 and 190 (on Child Labour), the ILO Encyclopaedia on Health and Safety and the Universal Declaration of Human Rights.

http://www.saiplatform.org/our-activities/potatoesvegetables/SAI_Platform_ Principles_Practices_Potatoes_Vegetables.pdf

---

[47] Unilever, 2002. Growing for the future II, Unilever and sustainable agriculture. Rotterdam, the Netherlands.

The Working Group on Fruits is chaired by Friesland Foods (RiedelDrinks), and the other members are Dole, Danone, McDonald's and Nestlé. Draft Principles and Practices for Sustainable Fruit Production were published in November 2005 and will be discussed with stakeholders and tested in pilot projects. http://www.saiplatform. org/our-activities/fruit/default.htm.

Market penetration
There is no certification system or label associated with the SAI-Platform. When the guidelines are finalized, implementation will be the responsibility of member companies. These members form a large part of the total market for food products. Suppliers to these firms may be asked to participate in pilot projects or in general to implement the guidelines.

Filière Qualité Carrefour

This quality supply chain label has been developed for cheese, meat, fruit and vegetables, and fish and seafood. The five key principles behind the programme are: taste and authenticity (traditional products typical for the region); long-term sustainable partnership along the entire chain; fair price; constant product quality; and environmental sustainability. After harvesting, no chemical treatments are applied for conservation. The specific "norms" are different for every supply chain. There is no information available about the verification system, costs and benefits.

Quality supply chains have been mainly developed in France, with 250 chains and more than 35 000 producers. However, there are also 350 quality supply chains with producers outside France and another 150 in development. Most of these are supplying local supermarkets owned by Carrefour. For example, there are 37 quality supply chains in Brazil supplying Brazilian supermarkets. For the French market the only quality supply chain relevant for fruit producers in developing countries is the one concerned with pineapple from Côte d'Ivoire.

Like Tesco Nature's Choice, this programme may be considered a business-to-business product specification and not a standard.

http://www.melchior.fr/melchior/melchior.nsf/0/95d4c3060c19cd61c1256e2b0034 4779?OpenDocument

Europe/Africa-Caribbean-Pacific Liaison Committee (COLEACP) Harmonized Framework

COLEACP is an interprofessional association of exporters, importers and other stakeholders in the EU-ACP[48] horticultural trade.[49] To improve market recognition of ACP produce and to respond to market demands for environmentally and socially responsible conditions of production, COLEACP took the initiative to encourage horticultural export associations to move towards harmonization of their Codes of Practice. The COLEACP Harmonized Framework is meant as a minimal set of food safety, environmental and social standards to be incorporated into national codes. There are 12 participating fresh-produce trade associations, coming from nine African and Caribbean countries.

The COLEACP Framework does not have any recognition in the US or EU markets. However, it does influence production practices through the capacity building within the participating export associations and the improvement of the codes of its members (for example, the Fresh Produce Exporters Association of Kenya, which developed the KenyaGap that has recently started a benchmarking process against EurepGAP).

---

[48] Country in Africa, the Caribbean or the Pacific that has signed the Cotonou agreement with the European Union.
[49] COLEACP, 2001. The COLEACP Harmonized Framework for ACP Codes of Practice. See: http://www.coleacp.org.

5.4. CIVIL SOCIETY AND NGO STANDARDS
Rainforest Alliance

The Sustainable Agriculture Network (SAN – formerly the Conservation Agriculture Network) is a coalition of ten conservation-oriented NGOs in the Americas. The programme initially focused on the environmental impact of production methods and on habitat conservation, but has increasingly incorporated standards for community relations and labour conditions. The Rainforest Alliance is the main force behind the initiative, and its Costa Rican office is the secretariat of SAN (the Rainforest Alliance head office is in New York).[50]

The product-specific standards have been developed together with producers, mainly from Costa Rica and other Latin American countries. The programme has set standards for five tropical crops: bananas, citrus, coffee, cocoa, and ferns and ornamental plants.

Currently "whole farm" standards are being developed for farms that grow additional crops for which no crop-specific standards currently exist. Additional chain-of-custody requirements are being developed to ensure that certified products and non-certified products are segregated from the moment they leave the farm until they reach the consumer.

Requirements

Environmental standards include the prohibition against clearing of primary forest; requirements for soil and water management and conservation and buffer zones; detailed requirements for the use, storage and transport of agrochemicals; integrated pest management; criteria for waste management and recycling; and requirements for a monitoring system. With respect to the social criteria, the certified company should respect all ILO conventions ratified by the country in which they operate. For those issues for which the ILO conventions are not ratified, the certification standards apply directly. These include requirements for a social policy and communication to workers; contracts and wages; no discrimination; no child labour below 14 and specific conditions for young and disabled workers and pregnant women; no forced labour; freedom of expression and the right to organize and collective bargaining; occupational health and safety; working hours; training; accommodation; and linkages with local communities.

There is no accreditation system. In most cases the certification is done by the local SAN member. The Rainforest Alliance operates the certification system in Costa Rica and Honduras and in countries where there is no SAN member. All auditors, including those from SAN members, are trained by the Sustainable Agriculture Programme secretariat, i.e. the Rainforest Alliance. The producers pay for the auditing and certification costs, depending on service delivered (e.g. actual auditor days and travel, etc.) plus a fixed amount per hectare.[51]

Penetration and markets

Promotion and uptake of the standards have mainly been confined to the Americas. The former Better Banana Project and ECO-OK seals have been replaced by the Rainforest Alliance Certified label. The seals are used mostly in public relations activities of certified producers, and in relations between producers and buyers (importers, wholesalers and retailers). The label is administered by the Rainforest Alliance, with a fee for use of the

---

[50] Other SAN members are Conservación y Desarrollo (CyD) in Ecuador; Centro Científico Tropical (CCT) in Costa Rica; Toledo Institute for Development and the Environment (TIDE) in Belize; SalvaNatura in El Salvador; Instituto Para la Cooperación y Autodesarrollo (ICADE) in Honduras; Fundación Interamericana de Investigación Tropical (FIIT) in Guatemala; Pronatura Chiapas in Mexico; Fundación Natura in Colombia; and Imaflora in Brazil.

[51] SAN, pers. comm.

label on products, although this fee may be waived Since many products from certified facilities are not labelled, there is no clear market segment for Rainforest Alliance Certified products. However, for bananas, market penetration can be discerned from the market share of the certified companies. In 2000 Chiquita achieved 100 percent certification of its company-owned farms in Latin America. In addition, 30 percent of the independent farms that supply Chiquita with bananas have achieved certification. Chiquita claims the production from these farms cover 90 percent of Chiquita's banana volume into Europe and approximately two thirds of the volume to North America. Reybanpac in Ecuador is also certified.

The Rainforest Alliance has recently upgraded its marketing efforts through the Certified Sustainable Products Alliance, funded by the United States Agency for International Development (USAID). It focuses on timber, coffee and bananas.

http://www.rainforestalliance.org/programs/agriculture/san/index.html

6. Environmental standards

In this chapter, environmental standards are discussed. Most standards in the previous chapter also have requirements related to the environment. The difference is that the standards in this chapter do not have requirements that are not related to the environment. (Standards for organic agriculture are not discussed here but in the next chapter.)

6.1. INTERNATIONAL AGREEMENTS
International codes and conventions related to pesticides
Maximum residue limits were discussed in Chapter 4 on food safety. In this paragraph agreements on pesticides are discussed that relate mostly to environmental consequences of pesticide production, distribution and use.

International Code of Conduct on the Distribution and Use of Pesticides
This Code is the worldwide guidance document on pesticide management for all public and private entities engaged in, or associated with, the distribution and use of pesticides. It was adopted for the first time in 1985 by the FAO Conference. Following the adoption of the Rotterdam Convention in 1998, it was updated and the revised version was approved in November 2002.

The Code is designed to provide standards of conduct and to serve as a point of reference in relation to sound pesticide management practices, in particular for government authorities and the pesticide industry. http://www.fao.org/ag/agp/agpp/ Pesticid/Default.htm

Rotterdam Convention on the Prior Informed Consent Procedure for Certain Hazardous Chemicals in International Trade
In 1989, FAO and UNEP jointly introduced the voluntary Prior Informed Consent (PIC) procedure. Chapter 19 of Agenda 21, adopted at the 1992 Rio Earth Summit, called for a legally binding instrument on the PIC procedure. This resulted in the adoption by the Convention of the Prior Informed Consent Procedure for Certain Hazardous Chemicals in International Trade in 1998 in Rotterdam. It entered into force on 24 February 2004.

The Convention enables the world to monitor and control trade in certain hazardous chemicals. Export of a chemical covered by the Convention can only take place with the prior informed consent of the importing party. The Convention covers chemicals from which the PIC Secretariat has received official notification by at least two Parties from different regions that these chemicals have been banned or severely restricted in these countries.. The Convention currently covers 24 pesticides, 6 severely hazardous pesticide formulations and 11 industrial chemicals.

Once a chemical is included in the PIC procedure, a decision guidance document (DGD) is circulated to importing countries. These countries are given nine months to prepare a response (to allow import of the chemical, not to allow import, or to allow import subject to specified conditions) or an interim response. Decisions by an importing country must be trade-neutral (i.e. apply equally to domestic production as well as to imports). The decisions of the importing countries are circulated, and exporting country Parties are obligated under the Convention to take appropriate measures to ensure that exporters within their jurisdiction comply with the decisions. http://www.pic.int/

Stockholm Convention on Persistent Organic Pollutants

The Stockholm Convention, which entered into force in May 2004, is a global treaty to protect human health and the environment from persistent organic pollutants (POPs). POPs are chemicals that remain intact in the environment for long periods, become widely distributed geographically, accumulate in the fatty tissue of living organisms and are toxic to humans and wildlife.

The Convention agrees to eliminate (with certain exceptions for specific uses) nine pesticides[52], to severely restrict the use of DDT, and to prevent the unintentional production and release of dioxins, furans, HCB and PCBs (in particular via waste burning). The 12 chemicals concerned are also known as the "dirty dozen".

Convention text: http://www.pops.int/documents/convtext/convtext_en.pdf

For more information see http://www.pops.int

Montreal Protocol and methyl bromide

Under the Montreal Protocol on Substances that Deplete the Ozone Layer, developed countries agreed to phase out completely by 1 January 2005 their controlled uses of methyl bromide, an effective fumigant and pesticide. Developing countries committed to a slower phase-out. Quarantine and pre-shipment uses do not fall under the 2005 phase-out. In 1997, the Parties recognized that for some uses there were no technically or economically feasible alternatives. Therefore, they have allowed "critical use exemptions".

http://www.unep.ch/ozone/Treaties_and_Ratification/2B_montreal_protocol.asp

ISO 14001

ISO/TC 207 is the ISO Technical Committee responsible for developing and maintaining the ISO 14000 series. TC 207 consists of business and government experts from 55 countries. The first standards of the series were published in 1996.[53] The standard that can be implemented by companies and against which companies can be certified is ISO 14001 Environmental management systems – Specification with guidance for use. ISO 14004 gives guidelines on principles, systems and supporting techniques for the implementation of environmental management systems, including guidance that goes beyond requirements of ISO 14001. Other standards in the ISO 14000 series are "tools" for implementing an environmental management system. These tools deal with environmental monitoring and auditing, labelling and product life cycle assessment.

As with ISO 9001, ISO 14001 certifications certify management systems and not products. Therefore, products can not be labelled as ISO 14001-certified. However, an indication that the firm producing the product is ISO 14001 certified is permitted (although the ISO logo may not be used).

Requirements

ISO 14001 has been written to support implementation of environmental management systems in many different types of organizations, including manufacturing and service companies, government agencies, associations and NGOs. Requirements for certification are the development of an environmental policy, including an implementation and communication plan, definition of responsibilities, staff training activities, documentation and monitoring. Apart from compliance being required with local (environmental) rules and legislation, the standard does not set specific performance targets. Instead, ISO 14001 aims at continuous improvement.

---

[52] aldrin, chlordane, dieldrin, endrin, heptachlor, hexachlorobenzene (HCB), polychlorinated biphenyls (PCBs), mirex, and toxaphene

[53] ISO, 1998. ISO 14000 – Meet the whole family! Geneva. Available at: http://www.iso.ch.

Penetration and markets

A growing number of farms are being certified against the ISO 14001 standard. More and more ISO 14001 certified firm claims can also be found on products. ISO 14001 is rapidly becoming a default certification for plantations. Managers of large production units often claim that ISO 14001 has been very useful for them in structuring their documentation, providing environmental management tools and, in some cases, reducing costs. Because there is no price premium and the certification can be costly and requires extensive documentation, ISO 14001 might be less attractive for smaller agricultural operations.

6.2.   NATIONAL REGULATIONS

A country search of the Pesticide Action Network (PAN) Pesticides database gives pesticide registrations and Prior Informed Consent information per country.

http://www.pesticideinfo.org/Search_Countries.jsp#Europe%20and%20 Central%20Asia

United States

The United States does not have any purely environmental regulation that is imposed on growers outside the United States or that has any impact on them.

The US green seal does not include food. The Canadian Environmental Choice scheme's only food product is coffee.

European Union

Germany has a strict recycling and packaging law. All packaging materials must be easily recyclable. There are also restrictions on the type of pallets allowed.

Furthermore, there are various lifecycle ecolabelling schemes in Europe[54] that are supported by governments, but they do not cover food.

6.3.   CORPORATE STANDARDS

UK Assured Produce and Tesco Nature's Choice

Assured Produce is a UK Integrated Crop Management scheme. Assured Produce-certified products (with the red tractor logo) are sold by all UK multiple retailers. The Assured Produce scheme requires its member suppliers and growers to undergo regular audits.

Nature's Choice is a programme of Tesco. Nature's Choice promotes the use of beneficial insects rather than chemicals to control pests, and encourages water and energy efficiency and recycling. Growers are asked to draw up a farm conservation plan, which guides them in protecting important wildlife and landscapes. Tesco has pilot schemes that monitor the effects of Nature's Choice. All UK growers of fresh fruit, vegetables, flowers and other ornamental plants have to meet Nature's Choice standards and obtain third-party verification in order to do business with Tesco. Tesco accepts the standards of Assured Produce as equivalent to Nature's Choice. Several certification bodies in the United Kingdom (e.g. ProCert) offer Nature's Choice certification. Tesco is introducing the guidelines to foreign suppliers, particularly in Spain and in the Netherlands, but also in Kenya, Australia and New Zealand.

Because Nature's Choice is specific to Tesco, it may be considered a business-to-business specification and not a standard. However, it is not a product specification in the traditional sense as it refers to production methods and not product characteristics.

For more information, see http://www.tescofarming.com/navigate.php?go=cop2.

Other corporate standards with environmental requirements have been discussed in the previous chapter on sustainable agriculture and GAP.

---

[54] Such as Community Eco-Label Award Scheme, the Nordic Swan, the Swedish Environmental Choice, the German Blue Angel and the French NF environnement.

6.4. CIVIL SOCIETY AND NGO STANDARDS

Many NGO standards for the agricultural sector address environmental issues, but all of them have additional social requirements and are therefore discussed in Chapter 5 (sustainable agriculture) or in Chapter 8 (labour and social standards) if the social requirements are the most important part of the standard. Furthermore, organic standards are discussed separately in the next chapter. As a consequence, there are no "purely environmental" NGO standards to be discussed here.

7. Organic agriculture standards

Introduction

Organic production refers to holistic management of the agro-ecosystem, emphasizing biological processes and minimizing the use of non-renewable resources. Although the terms "organic", "ecological" or "biological" have developed in Europe and North America to distinguish organic from conventional agriculture, many low-input traditional agriculture systems in other parts of the world are also de facto organic systems.

Requirements

Organic standards for plant production typically include: criteria for conversion periods; seeds and propagation material; maintenance of soil fertility through the use and recycling of organic materials; and pest, disease and weed control. The use of synthetic fertilizers and pesticides and of genetically engineered organisms is prohibited. There are also criteria for the admission and use of organic fertilizers and natural pesticides.

Market penetration

Global retail sales of organic products in 2004 were estimated at some US$28 billion, up from US$10 billion in 1997. The European market was estimated at US$14 billion in total organic sales in 2004. The largest organic food markets in Europe are Germany, the United Kingdom, France and Italy. Although their populations are smaller, Scandinavian countries (in particular Sweden and Denmark), Austria and Switzerland are also important as they have relatively high consumption of organic foods per capita. In addition, organic foods tend to fetch higher prices in these countries. Consumption per capita tends to be lower in South European countries such as Spain, Portugal and Greece. Consumption of organic foods in the Netherlands is relatively low but imports are substantial. This is due to the re-export trade to other European countries. Some of the major European suppliers of organic foods are Dutch.

Market penetration in Europe was very rapid in the late 1990s due to consumer concerns for the environment and the search for healthy foods. This tendency was further fuelled by several well-publicized food contamination scandals such as the BSE crises and dioxin contamination. Organic foods have been shown to contain no or lower quantities of pesticide residues than conventional products and a large number of European consumers associate them with healthy eating. At the peak of growth in the late 1990s and early 2000s, retail sales of organic foods rose by over 20 percent annually. However, the growth rate has decreased recently. It was estimated at between 8 and 15 percent annually depending on the country in 2004. It should be noted however that this reduced rate is still much higher than the growth of conventional food sales (estimated at between 2 and 3 percent per year in Western Europe).

There are several reasons for this decrease. The most obvious one is the price difference with conventional foods. Organic foods are on average 20 to 30 percent more expensive than their conventional equivalents, even though this difference tends to diminish. In addition, the small quantity of supply gives rise to inefficiencies in the marketing chain. The lack of regularity and consistency of deliveries is often quoted as a constraint. It is expected that these constraints will be gradually solved as production increases and scale economies are generated throughout the marketing chain. This should lead to a further decrease in the price of organic foods. Another factor that is contributing to the expansion of the organic market is the rising involvement of large-scale retailers. Many mainstream supermarket chains have announced that they intend to carry more organic products. Moreover, recent years have witnessed the emergence of supermarket chains specialized in organic and natural foods. Consequently, the market for organic foods is projected to grow at a steady rate in the short and medium term.[55]

In the United States, retail sales of organic food and beverages were about US$12 billion in 2004 according to media reports.[56] Sales have been growing at over 20 percent annually since the early 1990s. Contrary to the European market, the growth rate has not decreased in the past few years. Experts indicate that the adoption of a single national standard under the National Organic Program in 2001 has been instrumental in supporting this growth. Other reasons include the involvement of large-scale retailers, the rapid rise of specialized supermarket chains (such as Whole Foods) and the consumer search for healthy foods.

7.1.   INTERNATIONAL AGREEMENTS
Codex Alimentarius guidelines
Codex formulated guidelines for the production, processing, labelling (including claims) and marketing of organically produced vegetables. The guidelines were adopted in 1999 and revised in 2001 to include provisions for livestock and livestock products.

CAC/GL32 Guidelines for the Production, Processing, Labelling and Marketing of Organically Produced Foods. http://www.codexalimentarius.net/download/standards/360/CXG_032e.pdf

IFOAM, FAO, UNCTAD International Task Force
In an effort to harmonize existing organic guarantee systems, a task force was formed by IFOAM, FAO and UNCTAD. This International Task Force on Harmonization and Equivalence in Organic Agriculture started work in 2003, and serves as an open-ended platform for dialogue. It works on proposals related to mechanisms establishing equivalence of standards, regulations and conformity assessment systems for the consideration of governments, the Codex Alimentarius Commission and other relevant bodies.

7.2.   NATIONAL REGULATIONS
United States
The first organic regulations were adopted in the United States (in Oregon in 1974 and California in 1979). US-wide standards were developed by the USDA in 2002 as part of the National Organic Program (NOP).

The NOP requires that all products sold in the United States as "organic" be certified by a certification body or a state certification programme that is accredited by the USDA. Accredited bodies may be based in the United States or abroad. For certification bodies in the NOP system, ISO Guide 65 accreditation is voluntary.

Foreign accreditation agencies may also be recognized by USDA to perform NOP accreditations. Recognized accreditation programmes as of August 2005 were those of Denmark, New Zealand, Canada (Quebec, British Columbia) and the United Kingdom. Certification bodies in those countries may be assessed by their own government agency to determine if they fulfil NOP requirements. The USDA is also working with the Governments of Canada, Israel and Spain to recognize their ability to accredit for the NOP system.

Another option is recognition of equivalence of a foreign organic certification programme. In this case, specific NOP accreditation of certification bodies is no longer necessary. The USDA is currently working with India, Japan, Australia and

---

[55] This section is a summary of original market research that the FAO contracted to a consultant and data obtained from specialized media sources

[56] Eurofood 2005 and Kortbech-Olesen, R. 2003. The North American Market pp 11-13. In: Liu P. 2004. Production and export of organic fruit and vegetables in Asia. Proceedings of the seminar held in Bangkok, Thailand, 3-5 November 2003. FAO Commodities and trade paper 6. Rome. FAO/Earth Net Foundation/IFOAM

the European Union to determine whether their organic certification programmes are equivalent to the technical requirements and conformity assessment system of the NOP.

NOP came into force in October 2002, and currently (August 2005) has 99 accredited certification bodies, of which 43 are registered outside the United States and 11 from developing countries (see Table 2). This does not include the certification bodies accredited under the above-mentioned recognized accreditation programmes.

The use of the USDA organic logo is voluntary, provided the requirements for use are fulfilled. For more information and the text of the regulation, see:

http://www.ams.usda.gov/nop/NOP/NOPhome.html

In addition to the organic standards and certification systems, an international voluntary Code of Practice for Organic Trade has been developed by the IFOAM traders group and was launched in February 2003. The core of the Code of Practice consists of eight principles, including transparency and accountability of negotiations and equitable distribution of returns. Any organic trader can sign up for the code and participate in a continual self-assessment process.[57]

For more information on IFOAM, please see: http://www.ifoam.org/index.html

European Union

France was the first country to adopt an organic regulation (1985). EU Regulation 2092/91, covering the labelling of organic foods, was adopted in 1991. The regulation has been corrected and amended many times.

EU regulation EEC 2092/91 provides for national accreditation of certification bodies or certification by national authorities. Certification bodies are usually required to conform to European standard EN 45011 or ISO Guide 65, both of which are standards for the operation of certification systems. The organic guarantee system of countries outside the European Union may be recognized as being equivalent, and those countries appear on a "third-country" list. The list may specify production units or inspection bodies within the country for which equivalency is determined. Countries outside the European Union on this list are: Argentina, Australia, Costa Rica, Israel, New Zealand and Switzerland.

For imports from non-listed countries, importers may obtain an authorization from individual EU Member States for each imported product, so-called article 11(6) authorizations. The importer must show evidence that the product was produced and inspected according to rules equivalent to EU organic standards and was certified by a certification body that operates in compliance with ISO Guide 65. Since 2002, an original certificate must be sent with the goods. This transitional arrangement is expected to be extended until December 2006. It is also expected that before that time a new system will be developed for technical equivalency evaluations.[58]

The use of the European organic logo is voluntary, provided the requirements for use are fulfilled. A consolidated version of the regulation as per 1 May 2004 can be found at http://europa.eu.int/eur-lex/en/consleg/pdf/1991/en_1991R2092_do_001.pdf

For amendments after this date, the following page should be consulted:

http://europa.eu.int/smartapi/cgi/sga_doc?smartapi!celexapi!prod!CELEXnumdoc&lg=EN&numdoc=31991R2092&model=guicheti

The European Union has also prepared a brochure on the organic regulation, but it is from 2001 and some information is outdated:

http://europa.eu.int/comm/agriculture/qual/organic/brochure/abio_en.pdf

---

[57] Predefined report from http://europa.eu.int/comm/agriculture/ofis_public/pdf/r8_0000_en.pdf, generated August 2005

[58] Predefined report from http://europa.eu.int/comm/agriculture/ofis_public/pdf/r8_0000_en.pdf, generated August 2005.

For up-to-date information on article 11(6) authorizations, please see:
http://europa.eu.int/comm/agriculture/ofis_public/index.cfm

7.3. CORPORATE STANDARDS

There are countless organic labels, reflecting the many organic certification programmes developed by organic farmer associations and certification bodies. Many of the original internal control systems of organic farmer associations developed into certification programmes. With the implementation of the first regulations, certification bodies were split off from the associations to provide independent certification. Some of these initially small certification bodies have grown into big businesses.

These certification bodies often offer certification programmes against national standards, but they may also offer certification against their own standard. The latter standards may go beyond the minimum requirements of the national regulations and of the IFOAM basic standard (see below) or include areas that are not yet regulated (e.g. for organic textiles). Examples are the standards of Ecocert in France.

Other large certification companies not originating from the organic community have also included organic certification in the services they offer, but in most cases they have not developed their own standards.

Table 2 - NOP-accredited certification bodies in developing countries

Argencert SRL	Argentina	November 2002
Food Safety SA	Argentina	July 2003
LETIS SA	Argentina	December 2002
Organización Internacional Agropecuaria	Argentina	October 2002
Bolicert	Bolivia	March 2003
Instituto Biodinamico	Brazil	July 2002
Certificadora Chile Orgánico SA	Chile	August 2004
Biolatina	Colombia	April 2002
Eco-Logica	Costa Rica	July 2002
Mayacert SA	Guatemala	May 2003
ETKO- Ecological Farming Controlling Organization	Turkey	January 2003

7.4. NGO STANDARDS

International Federation of Organic Agricultural Movements (IFOAM) Basic Standards
Introduction

The International Federation of Organic Agriculture Movements (IFOAM) was founded in 1972. IFOAM is an umbrella organization with a global membership base of organic farmers associations, certification bodies, traders, consumers and advisory bodies. Its headquarters are in Bonn, Germany.

IFOAM formulated the first version of the IFOAM Basic Standards (IBS) in 1980 and has revised them regularly ever since, first every two years, now about every three years. The IBS therefore preceded all legislation. The IBS serves as a guideline, on the basis of which public and private standard-setting bodies can develop more specific organic standards. The IBS also has criteria for the functioning of internal control systems for group certification. These were later adopted by the European Union and incorporated into the EU legislation.

The International Organic Accreditation Service (IOAS) accredits certification bodies that have organic certification programmes that comply with IBS and the IFOAM Accreditation Criteria for certification bodies. Because the IBS is a generic standard, the IOAS requires that certification bodies elaborate some standards in more detail. As of June 2005 there were 31 accredited bodies, six of which were from developing countries

Table 3 - Approved inspection bodies based in developing countries inspecting the exporters involved in imports from third countries according to article 11(6)

Latin America	
Argencert	Argentina
Instituto Biodinamico	Brazil
Biolatina S.a.c.	Peru
Bio Latina	Colombia
Biologicos del Tropico	Colombia
Bolicert	Bolivia
Cenipae	Nicaragua
Certimex H. Escuala	Mexico
PROA	Chile
SGS de Peru	Peru
SGS Uruguay	Uruguay
Asia	
BCS Öko Garantie Gmbh T	Turkey
Ekolojiktarim Kontrol Organ. Ltd Sirteki	Turkey
IMO Turkey	Turkey
INAC Gmbh (International Nutrition and Agriculture Certification)	Turkey
NASAA	Sri Lanka
OFDC Organic Food Dev. Center	Tajikistan
Organic Agriculture Certification Thailand	Thailand
SGS India	India
SGS-SSTC Standard	China
SKAL	Turkey
Africa	
Bionoor	Algeria
Center of Organic Agriculture	Egypt
Egyptian ctre Organic Agric	Egypt

Source: European Parliament Legislative Observatory. Procedure file reference CNS/2005/0094 http://www2.europarl.eu.int/oeil/FindByProcnum.do?lang=2&procnum=CNS/2005/0094.

(Argencert and Organización Internacional Agropecuaria SA from Argentina, Bolicert from Bolivia, Instituto Biodinamico from Brazil, Organic Agriculture Certification from Thailand and Organic Food Development & Certification Center from China). In 1999, the IFOAM Accredited Certification Bodies (ACBs) signed a multilateral agreement to facilitate acceptance of products that were certified by an ACB. However, the agreement contains an additional requirements clause stating that those products should also comply with standards beyond the IBS, i.e. those that the "accepting" body might have in its own standards.

Requirements

During the last revision of IBS, the standards for ecosystem management were strengthened to include issues of landscape, contamination control and soil and water conservation. There are ongoing discussions about whether the standards should also include criteria for labour conditions and other social issues, to which only a general reference is currently made.[59] Processing, packaging and traceability standards usually include requirements to prevent mingling of conventional and organically produced products. They also include criteria for additives and processing aids. Because long travel distances contribute to the use of external inputs, it is debated whether criteria for local sourcing and means of transport should be developed (the "food miles" debate).

Market penetration

IFOAM is the largest organic association, with over 750 member organizations in 108 countries. IFOAM members represent a large part of all organic producers. The IFOAM accredited seal may appear on the product only as part of the logo of the certification body and in the body's own promotional material. Some US and EU buyers require organic certification from an IOAS-accredited programme.

In addition to the organic standards and certification systems, an international voluntary Code of Practice for Organic Trade has been developed by the IFOAM traders group and was launched in February 2003. The core of the Code of Practice consists of eight principles, including transparency and accountability of negotiations and equitable distribution of returns. Any organic trader can sign up for the code and participate in a continual self-assessment process.[60]

For more information on IFOAM, please see: http://www.ifoam.org/index.html.

---

[59] Schmid, O. 2002. Comparison of EU Regulation 2092/91, Codex Alimentarius Guidelines for organically Produced Food 1999/2001 and IFOAM basic standards 2000. pp. 12-18; Riddle, J. and Coody, L. 2002. Comparison of the EU and US organic regulations. Both in: G. Rundgren and W. Lockeretz (eds). IFOAM Conference on Organic Guarantee Systems; Reader. Tholey-Theley, Germany: IFOAM.

[60] IFOAM, 2003. IFOAM Code of Conduct for Organic Trade: Guidance document. See: www.ifoam.org.

8. Labour and social standards

8.1. INTERNATIONAL AGREEMENTS
International human rights
The United Nations High Commissioner of Human Rights is the principal UN official with responsibility for human rights and is accountable to the Secretary-General. The Office of the High Commissioner provides an overview of all international human rights law and UN bodies involved in human rights. http://www.ohchr.org/english/law/ and http://www.ohchr.org/english/bodies/

- Apart from the International Bill of Human Rights there are seven core international human rights treaties:
- International Convention on the Elimination of All Forms of Racial Discrimination.
- International Covenant on Civil and Political Rights.
- International Covenant on Economic, Social and Cultural Rights.
- Convention on the Elimination of All Forms of Discrimination against Women.
- Convention against Torture and Other Cruel, Inhuman or Degrading Treatment or Punishment.
- Convention on the Rights of the Child.
- International Convention on the Protection of the Rights of All Migrant Workers and Members of Their Families.

International Labour Organization (ILO) labour conventions
Labour rights are human rights applied to employment situations. The ILO is a UN specialized agency that seeks the promotion of social justice and internationally recognized human and labour rights. Unique in the UN system, the ILO is a tripartite organization, bringing together representatives of governments, employers and workers in its executive bodies.

The first six International Labour Conventions in the industry sector were adopted in 1919. Several years later the International Court of Justice declared that the ILO's domain also extended to international regulation of work conditions in the agricultural sector.

The ILO's fundamental human rights conventions are61:

No.29 Force Labour Convention, 1930 (168 ratifications)
http://www.ilo.org/ilolex/cgi-lex/convde.pl?C029
No. 105 Abolition of Forced labour Convention, 1957 (165 ratifications)
http://www.ilo.org/ilolex/cgi-lex/convde.pl?C105
No. 87 Freedom of Association and Protection of the Right to Organize Convention, 1948 (144 ratifications)
http://www.ilo.org/ilolex/cgi-lex/convde.pl?C087
No. 98 Right to Organize and Collective Bargaining Convention, 1949 (154 ratifications)
http://www.ilo.org/ilolex/cgi-lex/convde.pl?C098
No. 100 Equal Remuneration Convention, 1951 (162 ratifications)
http://www.ilo.org/ilolex/cgi-lex/convde.pl?C100
No. 111 Discrimination, Employment and Occupation Convention, 1958 (163 ratifications)
http://www.ilo.org/ilolex/cgi-lex/convde.pl?C111

---

61 Information on ratifications from: http://www.ilo.org/ilolex/english/docs/declworld.htm, consulted 2 September 2005.

No. 138 Minimum Age Convention, 1973 (141 ratifications)
http://www.ilo.org/ilolex/cgi-lex/convde.pl?C138
No. 182 Worst Forms of Child Labour Convention, 1999 (156 ratifications)
http://www.ilo.org/ilolex/cgi-lex/convde.pl?C182

Additional ILO conventions that are especially relevant for the agricultural sector are:

No. 99 Minimum Wage Fixing Machinery (Agriculture) Convention, 1951

The convention calls for the creation or maintenance of "adequate machinery whereby minimum rates of wages can be fixed for workers employed in agricultural undertakings and related occupations." It states that "the employers and workers concerned shall take part in the operation of the minimum wage fixing machinery, or be consulted or have the right to be heard...on a basis of complete equality." Ratified by 53 countries. http://www.ilo.org/ilolex/cgi-lex/convde.pl?C099

No. 101 Holidays with pay (Agriculture) Convention, 1952

Requires that workers employed in agricultural undertakings shall be granted an annual holiday with pay. Ratified by 46 countries. http://www.ilo.org/ilolex/cgi-lex/convde.pl?C101

No. 110 Plantations Convention, 1958

This convention deals with employment conditions of plantation workers. It is a very long and detailed Convention that covers conditions of work, contracts of employment, the official encouragement of collective bargaining, methods of wage payment, holidays with pay, weekly rest, maternity protection, accident compensation, freedom of association, labour inspection, housing and medical care. Ratified by 12 countries. http://www.ilo.org/ilolex/cgi-lex/convde.pl?C110

No. 129 Labour Inspection (Agriculture) Convention, 1969

This convention provides that ratifying governments shall maintain a system of labour inspection in agriculture. Ratified by 40 countries. http://www.ilo.org/ilolex/cgi-lex/convde.pl?C129

No. 141 Rural Workers' Organizations Convention, 1975

This convention defines rural workers as any person engaged in agriculture, handicrafts or a related occupation in a rural area, whether as a wage earner or a self-employed person, such as a tenant, sharecropper or small owner-occupier. But the Convention only applies to those tenants, sharecroppers or small owner-occupiers who derive their main income from agriculture, and who work the land themselves, with the help of only their family or occasional outside labour. The Convention reaffirms trade union rights for rural workers. It goes on to state that "it shall be an objective of national policy concerning rural development to facilitate the establishment and growth, on a voluntary basis, of strong and independent organizations of rural workers [...]." Ratified by 38 countries. http://www.ilo.org/ilolex/cgi-lex/convde.pl?C141

No. 184 Safety and Health in Agriculture Convention, 2001

The Convention covers preventive and protective measures regarding machinery safety, handling and transport of materials, chemicals management, animal handling, and the construction and maintenance of agricultural facilities. Other provisions address the needs of young workers, temporary and seasonal workers, and women workers before and after childbirth. Ratified by three countries. http://www.ilo.org/ilolex/cgi-lex/convde.pl?C184

For more information on ILO work and guidelines for agriculture:

http://www.ilo.org/public/english/dialogue/sector/sectors/agri.htm

In addition governments, employers' and workers' organizations, and multinational enterprises are recommended to observe the Tripartite Declaration of Principles concerning Multinational Enterprises and Social Policy. This Declaration, adopted in 1977 and revised in 2000, sets out principles in the fields of employment, training, conditions of work and life and industrial relations.

http://www.ilo.org/ilolex/english/iloquerymtn1.htm

The Global Compact
Introduction

The Global Compact is a personal initiative of the United Nations Secretary-General dating from 2000. It is aimed at encouraging business leaders to voluntarily promote and apply within their corporate domains ten principles relating to human rights, labour standards and the environment. At present, approximately 1 300 companies, most of them large transnational companies, from all continents have signed on to the Global Compact.

Requirements

The Global Compact's ten principles enjoy universal consensus and are derived from: The Universal Declaration of Human Rights; The International Labour Organization's Declaration on Fundamental Principles and Rights at Work; The Rio Declaration on Environment and Development; and The United Nations Convention Against Corruption.

Human Rights
Principle 1: Businesses should support and respect the protection of internationally proclaimed human rights; and
Principle 2: make sure that they are not complicit in human rights abuses.

Labour Standards
Principle 3: Businesses should uphold the freedom of association and the effective recognition of the right to collective bargaining;
Principle 4: the elimination of all forms of forced and compulsory labour;
Principle 5: the effective abolition of child labour; and
Principle 6: the elimination of discrimination in respect of employment and occupation.

Environment
Principle 7: Businesses should support a precautionary approach to environmental challenges;
Principle 8: undertake initiatives to promote greater environmental responsibility; and
Principle 9: encourage the development and dissemination of environmentally friendly technologies.

Anti-corruption
Principle 10: Businesses should work against all forms of corruption, including extortion and bribery.

Market penetration

The Global Compact includes nearly 2 200 companies from more than 80 countries. There are 38 participating companies from the agricultural sector (primary production), of which 35 are from developing countries. There are a further 163 companies from the food and drinks industry, among them Danone and Nestlé subsidiaries in Argentina and South Africa.

The so-called Communication on Progress policy requires that participants develop an annual disclosure to their stakeholders on implementation actions within two years of joining the Global Compact initiative. The policy went into effect on 30 June 2005 for the 977 participants that have been in the Global Compact for at least two years. Of these companies, only 38 percent – or 367 companies – have developed Communications on Progress for their stakeholders. The rest either did not develop a Communication on Progress by the deadline or did not inform the Global Compact Office that such a communication had been developed.

http://www.unglobalcompact.org/Portal/Default.asp?

Norms on responsibilities of business

The Sub-Commission on the Promotion and Protection of Human Rights adopted in August 2003 the (draft) Norms on the responsibilities of transnational corporations and other business enterprises with regard to human rights. They have been transmitted to the Commission on Human Rights for consideration and eventual adoption. The norms were discussed in the last Commission meeting in April 2005, but views about the norms were still divided. They will probably not be adopted in the current version but may be amended and adopted in the future.

The Norms document is therefore not a formal treaty under international law.

http://www.unhchr.ch/huridocda/huridoca.nsf/(Symbol)/E.CN.4.Sub.2.2003.12.Rev.2.En?Opendocument

Organisation for Economic Co-operation and Development (OECD) guidelines

The OECD has also developed recommendations for multinational enterprises.
OECD Guidelines for Multinational Enterprises (1976, revised 2000)

These guidelines provide voluntary principles and standards for responsible business conduct in a variety of areas, including employment and industrial relations, human rights, environment, information disclosure, combating bribery, consumer interests, science and technology, competition, and taxation. http://www.oecd.org/department/0, 2688,en_2649_34889_1_1_1_1_1,00.html

8.2. NATIONAL REGULATIONS

The United States and Europe have their own national labour laws, but they are applicable only domestically. Some national regulations may have an impact on producers in developing countries.

United States

In the FDA Guide to Minimize Microbial Food Safety Hazards for Fresh Fruits and Vegetables of 1998, which some foreign suppliers are required to implement (see Chapters D2 and D3), reference is made to the Occupational Safety and Health Act. Regulations under this Act have Title 29 CFR codes. http://www.osha.gov/index.html

The referenced regulations in the FDA guide are:
- 29 CFR 1928.110 subpart I, on field sanitation
- 29 CFR 1910.141, subpart J, on sanitation for enclosed packing facilities

European Union

Article 50 of the Cotonou Agreement of 2003 states that the European Union and the ACP countries will respect the ILO's Core Labour Standards.

The Belgian social label

In 2002 the Belgian Parliament adopted the Law for the promotion of socially responsible production. The law aims to create a label that companies voluntary affix to their products if the eight core ILO labour conventions (see chapter 8.1 above) are respected during the production process.[62] In April 2003 the official decisions necessary to implement the label were taken. The first labels have already been granted. Companies with SA8000 certification may use the Belgium social label.

To be granted use of the label, a company must submit a preliminary request to the implementing committee. The committee decides whether the company is eligible. If yes, the company requests one of the accredited auditing bodies to review the documentation and execute an audit. On the basis of the preliminary report, the company may take immediate corrective action if necessary. The auditing body

---

[62] http://www.just.fgov.be/mopdf/2002/03/26_2.pdf.

submits a final report to the committee, which advises the minister on the granting of the label.

To be accredited, auditing bodies have to be accredited against EN 45004 (general criteria for the operation of inspection bodies) or be SAI-accredited to certify against SA8000. To date, only SAI-accredited bodies are accredited for auditing against the Belgium social label.

More information may be obtained from http://www.social-label.be/.

Italy and Denmark are developing a similar social label. The European Union has decided to wait until more countries have implemented a social label before a decision is taken on whether to develop a European label.

8.3.   CORPORATE STANDARDS
Occupation Health and Safety Assessment Series (OHSAS) 18001
OHSAS 18001 was created by a group of 14 standards bodies, certification bodies and specialist consultancies.[63]

OHSAS 18001 is an Occupational Health and Safety Assessment Series for health and safety management systems. A main driver for the series was to try to remove confusion in the workplace due to the proliferation of certifiable OHSAS specifications. A number of older documents were used in the creation process, some of which were standards and certification programmes offered by the participating bodies.

OHSAS 18001 has been developed to be compatible with the ISO 9001 and ISO 14001 management systems standards. As a management systems standard, it does not state specific performance criteria, nor does it give detailed specifications for the design of a management system. It does require companies to have a health and safety manual, and for convenience sells a blueprint.

OHSAS forms part of the Health and Safety Electronic Book, published by the British Standards Institution. This includes not only the text from OHSAS 18001, but also BS8800 and various other materials and information.

No information on market penetration was found.

http://www.ohsas-18001-occupational-health-and-safety.com/index.htm

Blueprint manual available at: http://www.osha-occupational-health-and-safety.com/

8.4.   CIVIL SOCIETY AND NGO STANDARDS
International Confederation of Free Trade Unions (ICFTU)/International Trade Secretariats (ITS) Basic Code of Labour Practice
The ICFTU was established in 1949 and has 231 affiliated organizations in 150 countries, with a membership of 158 million. It is a confederation of national trade union centres, each of which links together trade unions of that particular country. It also maintains close links with Global Union Federations, which link together national unions from a particular trade or industry at international level. The agriculture sector is included in the International Union of Food, Agricultural, Hotel, Restaurant, Catering, Tobacco and Allied Workers Association (IUF).

The Basic Code of Labour Practice was developed in by the ICFTU/ITS Working Party on Multinational Companies, in consultations with trade union organizations and other stakeholders. It was adopted by the ICFTU Executive Board in 1997. The Basic Code aims to establish a minimum list of standards to be included in all codes of conduct covering labour practices. These could be company codes of

---

[63] National Standards Authority of Ireland; Standards Australia; South African Bureau of Standards; British Standards Institution; Bureau Veritas Quality International; Det Norske Veritas; Lloyds Register Quality Assurance; National Quality Assurance; SFS Certification; SGS Yarsley International Certification Services; Asociación Española de Normalización y Certificación; International Safety Management Organisation Ltd; Standards and Industry Research Institute of Malaysia; International Certification Services.

conduct, especially codes that are meant to apply to the international operations of a multinational company.

The Basic Code promotes the primacy of international labour standards and to incorporate freedom of association and the right to collective bargaining. It is not intended that collective bargaining agreements be limited to the provisions of the code. The Basic Code is also meant to encourage the use of consistent language in codes of conduct and to assist any trade union in negotiations with companies and in working with NGOs in campaigns involving codes of conduct.

The content of the code follows the core ILO conventions (29, 87, 98, 100, 105, 111 and 138) plus the Workers Representatives' Convention (No. 135) of 1971. The ICFTU Basic Code is a generic code and not meant for certification purposes.

http://www.icftu.org/displaydocument.asp?Index=991209513&Language=EN

Fair-trade

Introduction

The fair-trade initiatives seek to provide better market access and trading conditions for small-scale farmers and to improve the conditions for workers. The Fairtrade Labelling Organizations (FLO) International was founded in 1997 as an umbrella organization of 19 national fair-trade labelling initiatives. The Fair Trade Association of Australia and New Zealand and Comercio Justo Mexico joined FLO recently. Producers and traders are represented on the board and on various committees.

Since January 2003, the certification unit has been a legally independent certification body and has achieved ISO Guide 65 accreditation. Inspection is done by local auditors, while the certification decision is taken at FLO headquarters in Bonn, Germany. FLO will also recognize Certimex, an organic certification body in Mexico, as an inspection body for FLO.

Certification had been free of cost to producers, but since December 2003 they have had to pay a certification fee. Initial fees depend on the size of the operation, whereas renewal fees depend on the value of products sold under the fair-trade arrangements.

Requirements

FLO has developed one set of generic standards for small producers and one set of generic standards for hired labour. They contain standards for social development, economic development, environmental development and labour standard (if applicable).

Standards for farmers' associations and cooperatives set criteria for a democratic and participative organizational structure. Labour standards for plantations and factories include criteria for: freedom of association; wages and accommodation; and occupational health and safety. No child or forced labour can occur.

Traders who wish to trade in FLO-labelled products have to register and respect the trading standards. They must pay the FLO minimum price plus the fair-trade premium, partially pay in advance when producers ask for it, and commit themselves to a long-term trade relationship. To determine the minimum price, FLO first defines, per product and per region, the Cost of sustainable Production (CoP), and Cost of sustainable Living (CoL). The minimum price covers at least the CoP and CoL. For the registration of traders, FLO has developed a trader application evaluation policy. Registered traders have to report their sales figures to FLO; the figures are checked against reported sales figures from producer organizations.

In addition to the generic standards there are product-specific standards, with separate versions for small farmers and hired labour situations. Currently, product-specific standards for small farmers exist for bananas, cocoa, coffee, cotton, dried fruit, fresh fruit and vegetables, honey, juices, nuts and oilseeds, quinoa, rice, spices and herbs, sugar, tea and wine. For hired labour situations, standards exist for bananas, cut flowers, fresh fruit, juices, ornamental plants, sport balls, tea and wine. There are also "contract production standards" for cotton from India and Pakistan, and rice from India.

New environmental standards for both the generic standards and the product-specific standards are being developed and will probably enter into force in 2006.

Penetration and markets

Since 2003, the various national fair-trade labels have been replaced by the new International Fairtrade Certification Mark. In the United States, Canada, Mexico and Switzerland the national labels will continue to be used for the time being. The international label will greatly reduce logistical costs, as products will not have to be packaged separately for each destination.

For the last several years, year-on-year growth of fair-trade volumes has been around 20 percent. The largest volumes have been reached for bananas, with some 100 000 tonnes traded in 2004. The largest fair-trade markets are Switzerland and the United Kingdom. In 2004, sales volumes of other fresh fruits were estimated at 5 000 tonnes, of which 2 150 were for pineapples. Other certified fruits included mangoes, oranges, lemons, applies and grapes.

For more information on FLO, the standards and markets, see:
http://www.fairtrade.net/index.html
For more information on the certification system and FLO-Cert Ltd, see:
http://www.fairtrade.net/sites/certification/certification.html
For the trader evaluation policy, see: http://www.fairtrade.net/pdf/FLO%20 Trader%20Application%20Evaluation%20Policy.pdf

Social Accountability (SA) 8000 Standard

Introduction

SA 8000 is a workplace standard developed by Social Accountability International (SAI) in 1998. The SAI Advisory Board includes experts from trade unions, businesses and NGOs from various countries. SAI is based in New York.

The standard promotes the implementation of International Labour Organization (ILO) conventions covering social justice and working conditions. The standards were initially developed for the manufacturing industry, and approved for use in the agriculture sector in 2000. The standards were revised in 2001 and the guidance documents in 2004. They can be purchased for US$100 (US$35 for small businesses and NGOs).

SAI accredits certification bodies to audit production facilities. Accreditation requirements include demonstrated adherence to ISO/IEC Guide 62 and indications as to how the necessary information to make a certification decision will be obtained. The individual auditors performing the inspections must be trained in SA000 and be accredited as well. SAI is currently exploring whether and how NGOs that have experience in promoting better labour conditions may become accredited.

Companies that do a substantial amount of sourcing from contracted suppliers can join the Corporate Involvement Programme as an Explorer (level 1) or a Signatory (level 2). Signatories work with SAI to develop and execute a plan for implementation of SA8000 and/or certification of designated company-owned and/or supplier facilities within agreed-upon scope. They have to communicate progress on SA8000 implementation goals and progress to stakeholders via SAI verified annual reports or the SAI Multi-stakeholder evaluation and recommendation process. Signatories may use SAI and SA8000 logos in their communications.

Requirements

The requirements for SA8000 certification can be summarized as follows (for details please see the original standard and guidance documents, which can be found at the website provided below):

- No workers under the age of 15; minimum age lowered to 14 for countries operating under the ILO Convention 138 developing-country exception; remediation of any child found to be working.
- No forced labour, including prison or debt bondage labour; no lodging of deposits or identity papers by employers or outside recruiters.
- Provide a safe and healthy work environment; take steps to prevent injuries; regular training for workers in health and safety; system to detect threats to health and safety; access to bathrooms and potable water.
- Respect the right to form and join trade unions and bargain collectively; where law prohibits these freedoms, facilitate parallel means of association and bargaining.
- No discrimination based on race, caste, origin, religion, disability, gender, sexual orientation, union or political affiliation, or age; no sexual harassment.
- No corporal punishment, mental or physical coercion or verbal abuse.
- Comply with the applicable law on working hours but, in any event, no more than 48 hours per week with at least one day off for every seven-day period; voluntary overtime paid at a premium rate and not to exceed 12 hours per week on a regular basis; overtime may be mandatory if part of a collective bargaining agreement.
- Wages paid for a standard work week must meet the legal and industry standards and be sufficient to meet the basic need of workers and their families; no disciplinary deductions.
- Facilities have to integrate the standard into their management systems and practices.

Penetration and markets

The SAI-SA8000 label is not used on products. Certified facilities and signatories may use the label in their communications. Consequently there is no differentiated market for SA8000-certified products. The "market" could be understood to be those large buyers (e.g. retailers) that are Signatory Members and consequently try to move their suppliers towards SA8000 certification.

SA8000 was approved for use in the agriculture sector in 2000, and so far 25 agriculture facilities/divisions have been certified, covering growing, packing and processing of bananas, pineapples, canned fruit, coffee, tobacco and wine. One certified facility may include several farms, as is the case of Chiquita and Dole certifications. Certifications in fruit and vegetable production are provided in Table 4.

Dole has been a signatory member since 1999. Initially the progress in terms of certification was very slow, but according to the SAI web site 61 farms have now been certified.

SA8000 standard: http://www.sa-intl.org/Document%20Center/2001StdEnglishFinal.doc

For more information on SAI and SA8000 related activities: http://www.sa-intl.org

Ethical Trading Initiative (ETI)

Introduction

The ETI is a multi-stakeholder alliance in the United Kingdom. It has a tripartite structure in which NGOs, unions and the business sector are represented, with support from government. The ETI focuses on ethical sourcing by companies, in particular retail chains. Although the ETI is a national initiative, the sourcing of its members and therefore its impact are international. The ETI is a learning initiative to gain insight into how social standards can be developed and implemented.

Requirements

The Base Code was first published in 1998. It is based on the eight core ILO conventions. Furthermore it draws on conventions: 175 Part-time Work Convention of 1994; 183 Maternity Protection Convention of 2000; 177 Home Work Convention of 1996 and its related recommendation; and 159 Vocational Rehabilitation and Employment (Disabled Persons) Convention, 1983 and its related recommendation. The ETI published smallholder guidelines in 2005.

The ETI conducts various pilot projects to learn about implementing the Base Code in various circumstances. Pilot projects conducted to date in the agriculture sector are a horticulture project in Zimbabwe and a project in the wine industry in South Africa. A pilot project on bananas in Costa Rica was stopped at mid-term due to the inability of the three parties of the Costa Rican tripartite steering committee to agree.

The horticulture pilot project in Zimbabwe resulted in the formation of the Agricultural Ethics Assurance Association of Zimbabwe, a tripartite association of local business, trade unions and development agencies. They developed their own code of practice , which is expected to gain recognition of equivalence by the Dutch MSP flower code.

Penetration and market

Companies involved in the ETI execute internal business evaluation programmes to assess compliance with the ETI Base Code and subsequently try to address any non-conformities encountered in the evaluations. There is no certification system and consequently no label or specific market. There is no information available on extent to which ETI members request implementation of the Base Code from their suppliers, or if and how they monitor this.

Large food retailer members of ETI are: Asda, Marks and Spencer, Sainsbury's, Somerfield, Tesco and the Co-operative group (Co-op UK).

Other members from the fruit and vegetable sector are: Chiquita International Brands, Fyffes Group. http://www.ethicaltrade.org/

The Base Code and principles of implementation can be downloaded from:
http://www.ethicaltrade.org/Z/lib/base/index.shtml

Table 4 - SA8000 certifications in the fruit and vegetable sector, March 2005

Company	Country	Product	Sites
Chiquita Brands	Costa Rica	Bananas	
Standard Fruit Company de Costa Rica S.A.	Costa Rica	Bananas and pineapples	10 sites
Cirio del Monte Hellas S.A.	Greece	Canned fruit	
Compañia Bananera Guatemalteca Ind. S.A. / Operación Compra de fruta Costa Sur.	Guatemala	Fruit purchase	Offices; central office and warehouse at Tiquisate
Chiquita Division Maya Guatemala/Honduras	Guatemala / Honduras	Bananas	Guatemala: 9 sites Honduras: 18 sites
Tata Coffee Limited (Plantation Division)	India	Coffee, tea, pepper, cardamom, vanilla	Coorg, Hassan, Chickmagalur Districts
PT Great Giant Pineapple	Indonesia	Pineapple	
Mehadrin Tnuport Export	Israel	Fresh fruits	
Yevuley Nanir (R.A.C.S)	Israel	Vegetables, fruit, bulbs, flowers	
Cirio del Monte Kenya Ltd.	Kenya	Pineapples, other food	
Dole Philippines, Inc.	Philippines	Pineapples, other products	
Stanfilco, A Division of Dole Philippines, Inc.	Philippines	Bananas	6 sites
Bocas Fruit Company	Panama	Bananas	all operations

Source: SAI website, consulted July 2005

Annex 1 - Matrix of standards (logistical and general labelling standards and single-company standards have not been included)

	Intergovernmental	ISO	National regulations (that affect foreign producers)	Corporate	Civil society/ NGOs
Framework	GATT, TBT, SPS, TRIPS, GATS	Guides 2, 7, 28, 53, 59, 60, 67, 68, 10011, 17000series	EN 45011, EN 45012	AA1000 and GRI (with civil society)	ISEAL code of good practice for setting social and environmental standards
Quality	Codex commodity standards	ISO 9000	USDA grades, USFQAP, EU marketing standards	Qualipom'fel	n.e.
Phytoosanitary	IPPC (ISPM 7 on export certification system and ISPM 15 on wood packaging)	n.e.	US Title 7 CFR 319.56 (import permits) and CFR 319.40 (wood packaging) EU Directive 2000/29/EC (phytosanitary certificates) and 2004/102/EC (wood packaging)	n.e.	n.e.
Food safety	Codex texts, incl.: food hygiene (incl. HACCP), pesticide MRLs, MLs for additives, contaminants, irradiation, food safety assessment of GMOs	ISO 22000	US Food Code, FDA guide for ff&v 1998, MRLs (tolerances), irradiation. EU Food Law, directive on food hygiene, MRLs, irradiation	GFSI; BRC, IFS, Dutch HACCP, SQF, Certiclean, EFSIS, AIB gold standard certification against FDA guidelines, MDI	n.e.s
Purely environmental (excl. organic)	Pesticide Code, PIC, Stockholm convention Montreal protocol,	ISO 14001	Not relevant*	Not relevant**	Not relevant***
Sustainable agriculture, GAP	Agenda 21, SARD, GAP	n.e.	Not relevant*	EurepGAP, SAI-Platform, Filière Qualité Carrefour, COLEACP Framework	Rainforest Alliance
Organic	Codex, ITF	n.e.	US NOP EU regulation 2092/91	Certification body's own standards	IFOAM
Social	Human rights treaties, ILO, Global compact, OECD guidelines for multinationals	Future ISO CSR standard	Belgian social label	OHSAS 18001	ICFTU/ITS Basic Code of Labour Practice, Fair trade, SA 8000, ETI

n.e. = not existing

* National domestic regulations on sustainable agriculture and the environment do not apply to foreign producers and exporters and therefore are not dealt with in this report.

** National domestic regulations on sustainable agriculture and the environment do not apply to foreign producers and exporters and therefore are not dealt with in this report. Corporate environmental codes that affect fruit and vegetable exports from developing countries to the United States and the European Union. Corporate environmental codes such as waste management are implemented in the country of operation and do not affect suppliers and are furthermore often "single company" standards. ***There are no purely environmental NGO codes that are relevant for the fruit and vegetable sector, other than organic standards. The Rainforest Alliance standards initially were a purely environmental standard but the Alliance has incorporated many social standards and therefore is classified under "sustainable agriculture".

Part 2

Overview of existing analytical work on the impacts of private standards on trade

Introduction

This chapter presents an overview of existing analytical work related to private standards and trade, with a focus on fruit and vegetable exports from developing countries.

The abstracts provided in this chapter are not summaries, but they rather highlight the content that is most relevant to this topic. Only those case studies that relate to the fruit and vegetable export sector have been included. The overview focuses on private standards, therefore it is far from exhaustive regarding studies on specific import regulations. The publications on regulations included in this overview have relevance for the analysis of the impacts of private standards, for example because of an interesting methodological approach.

For the sake of convenience and in order to give an indication of what aspects have been covered by research, the studies have been sorted by subject headings. However, a paper may address several subjects. In this case the paper has been classified under the subject that is most prominently addressed (to the personal judgement of the author of this overview). Only in rare cases has a paper been mentioned twice.

The overview is far from exhaustive. Many more studies exist, especially on the market aspects (notably on market volumes for labelled products, on consumer preferences and on consumer willingness-to-pay).

1. Extensive research programmes covering various aspects of private standards

Challenges and Opportunities Associated with International Agro-Food Standards (World Bank)

This World Bank research programme was launched in 2002 and the final report was published in 2005. The objective of the research programme is to improve the understanding of the development community regarding the challenges and opportunities for developing country trade associated with rising international food safety and agricultural health standards.

Contact: http://web.worldbank.org/WBSITE/EXTERNAL/TOPICS/TRADE/0,,contentMDK:20334931~menuPK:634021~pagePK:148956~piPK:216618~theSitePK:239071,00.html

Coordinator: Steven Jaffee sjaffee@worldbank.org

General Papers

1. Jaffee, S. and S. Henson. 2005. Food Safety and Agricultural Health Standards: Challenges and Opportunities for Developing Country Exports. Synthesis report.

Abstract: Strategies: exit, voice and compliance. Costs and benefits of (non) compliance. Distributional effects of standards. Implications for technical assistance.

2. Jaffee, S. and S Henson. 2004. Standards and Agro-Food Exports from Developing Countries: Rebalancing the Debate. World Bank Policy Research Working Paper 3348, June 2004.

Keywords: SPS standards; capacity-building; compliance costs; high-value agro-food trade.

Abstract: Rising standards serve to accentuate underlying supply chain strengths and weaknesses and thus impact differently on the competitive position of individual countries and distinct market participants. Some countries and/or industries are even using high quality and safety standards to successfully (re-)position themselves in competitive global markets.

3. Sewadeh, M. and V. Ferrer. 2004. Donor Support for SPS Capacity Building: taking stock and drawing lessons. Report prepared for the World Bank's International Trade Department. Washington D.C.

Abstract: Most support for capacity building in SPS issues is provided at times of (looming) trade disruptions, when options and acceptable time frame for change are reduced. Bilateral assistance is driven by self-interests for safe food imports. Most projects are sector specific and give no attention to other trade issues. There is little collaboration among agencies. Many projects target either the public or private sector. Partners in the private sector are often the "best" firms, which is cost-effective but re-enforces uneven distributional impact patterns. SPS management should become a mainstream element in strengthening competitiveness and technical assistance should aim at maximizing the strategic options of developing countries.

4. Henson, S., S. Jaffee, C. de Haan and K. van der Meer. 2002.

Sanitary and Phytosanitary Requirements and Developing Country Agro-Food Exports: Methodological Guidelines for Country and Product Assessments.

Keywords: SPS capacity; compliance strategies; country assessment; interview guide.

Abstract: A framework for assessing levels of sanitary and phytosanitary (SPS) management capacity in developing countries and implications for major agricultural and food exports. Considering the strategies adopted to comply with SPS measures in export markets and the associated costs of compliance. Including interview guides.

5. Wilson, J. S. and V. O. Abiola (eds.). 2003. Standards and Global trade: A Voice for Africa.

Keywords: Kenya, Mozambique, Nigeria, South Africa, Uganda.

Abstract: Identifies the specific capacity constraints, opportunities, and institutional reform needed for market-access in five African countries (all five case studies include the horticulture sector). Places trade facilitation measures and standards (both voluntary and mandatory technical standards) within a broader developmental context.

Contact: http://www1.worldbank.org/publications/pdfs/15473frontmat.pdf

Selected case studies producer countries

6. Aloui, O., and L. Kenny. 2004. Case Study on Cost of Compliance to SPS Standards for Moroccan Exports.

Abstract: Focus on tomato and citrus exports. The Moroccan national pesticide registration should be updated. At the farm level, compliance costs with the EurepGAP standard form 8 percent of the total farm gate costs. Small farms require a longer implementation time. For citrus growers, the requirement of mobile sanitation facilities will result in higher costs. At the packing house level, HACCP is fairly new but has already been integrated in several management strategies. Many managers have noted benefits from ISO 9001 implementation: better worker performance, greater efficiency, less rejects and easy tracing of the source of problems. The residue testing required by the British Retail Consortium (BRC) can not be done in the government laboratory and samples must be sent to Europe. Differences among standards are the most serious problem.

7. Canale, F. 2004. The Phytosanitary Capacity of Developing Countries. (not online)

8. Henson, S. and S. Jaffee. 2004. Jamaica's Trade in Ethnic Foods and Other Niche Products: the Impact of Food Safety and Plant Health Standards.

Abstract: Core SPS management framework and human capital are available, but better coordination of SPS controls is needed. Investment has been reactive, and overall private investment in SPS control systems has been very limited. Other competitiveness problems limit economic benefits of enhanced trade-related SPS controls unless broader efforts to enhance export competitiveness are undertaken. Exporters can also choose to expand sales in value-added processed food products.

9. Jaffee, S. 2003. From Challenge to Opportunity: Transforming Kenya's Fresh Vegetable Trade in the Context of Emerging Food Safety and other Standards in Europe. World Bank, ARD Discussion Paper No. 2. Washington, D.C.

Abstract: The Kenyan fresh produce industry—with the assistance of the Government of Kenya and others—is effectively and proactively complying with rising standards and uses this as a competitive advantage. Remaining challenges are the documentation of safety of fresh produce sourced from smallholders, upgrading of small and medium enterprises and recognition by the European Union of the Kenya Plant Health Inspectorate Service as a "competent authority".

10. Jaffee, S. 2004. Delivering and Taking the Heat: Indian Spices and Evolving Product and Process Standards.

Abstract: Indian spice trade has earned a reputation for product quality and marketing service. Recent changes in regulatory and commercial requirements in some markets have triggered responses in production, post-harvest, and processing practices; in quality assurance and supply chain management systems; and in monitoring and testing products. Via effective private and public sector collaboration India also influences the "rules of the game" internationally. India is fully expected to meet remaining and emerging commercial and regulatory challenges.

11. Manarungsan, S., J. Naewbanij, and T. Rerngjakrabhet. 2004. Costs of Compliance to SPS Standards: Thailand Case Studies of Shrimp, Fresh Asparagus, and Frozen Green Soybeans.

Abstract (asparagus only): Over the last five years asparagus exports to Japan have grown strongly and exporters expanded into new markets. The tightening of residue limits in Japan pressured farmers to shift to organic farming or reduced chemical usage, increasing production costs by 165 percent and lowering yields by 20 percent. But they received a 29 percent price premium over conventional asparagus. For exporters, the cost of compliance is 100 percent higher than before (63 percent laboratory analysis, 37 percent quality systems). The government has invested in: laboratories; standards and guidance for Good Agricultural Practices; pesticide test kits; high-yielding and pest-resistant varieties; and technology to prepare and preserve natural pesticides.

12. Mbaye, A. A. 2004. Sanitary and Phytosanitary Requirements and Developing Country Agrifood Exports: An Assessment of the Senegalese Groundnut Sub-Sector.

Abstract (edible groundnuts only): Recently exports of edible groundnuts dropped by 90 percent as a result of the decline in output and yields in the entire groundnut sector. On the trade side, the main difficulty is meeting aflatoxin standards. A cost/benefit analysis of meeting this standard found a net benefit of 92 billion Franc-CFA for edible groundnuts through higher prices and the possibility of selling greater quantities.

Selected Buyers Surveys

13. Lamb, J., J. Velez, and R. Barclay. 2004. The Challenge of Compliance with SPS and Other Standards Associated with the Export of Shrimp and Selected Fresh Produce Items to the United States Market.
Abstract: Overview of applicable standards and main SPS issues encountered, including the history of contaminations and entry refusals. The section on fresh produce focuses on raspberries, cantaloupe, asparagus, snow peas and mango. Analyses the strategies of suppliers from the buyers' perception and the cost of non-compliance (including refused shipments at the port etc.)

14. Willems, S., E. Roth and J. van Roekel. 2004. Changing European Public and Private Food Safety and Quality Requirements: Challenges for Developing Country Fresh Produce and Fish Exporters. European Union Buyers Survey. The World Bank, Rural Development Department. Washington, D.C.
Abstract: Transnational companies guarantee quality and safety under a private label in a vertically integrated chain. In collaborative chains, buyers support suppliers to implement standards. In transaction-oriented chains with intermediaries, suppliers are not regularly informed about standards and run the risk of non-compliance. The reasons for changing suppliers include: food safety; volume and reliability of supply; price; quality and packaging; social and ethical issues and political conditions.

Training

Training seminar for World Bank staff, January 27-28, 2004
     15. Bureau, J-C. 2004. Raising the Bar on Product and Process Standards: Economic Principles. World Bank, Washington DC.

Articles in journals

16. Jaffee, S. and O. Masakure. 2005. Strategic use of private standards to enhance international competitiveness: Vegetable exports from Kenya and elsewhere. In. Food Policy Vol. 30, nr.3, pages 316-333. Elsevier.
     Keywords: Private standards; Food safety; Brand reputation; Supply chain restructuring.
     Abstract: Leading Kenyan fresh produce suppliers have re-positioned themselves at the high end segments of the market – those most demanding in terms of quality assurance and food safety systems. Factors having influenced this positioning include: relatively high freight costs and low labour costs, the emergence of more effective competition in mainstream product lines, and strong relationships with selected retail chains.
     Contact: sjaffee@worldbank.org, aar00om@rdg.ac.uk

Regoverning Markets (IIED)

Securing Small-scale Producer Participation in Restructured Agri-food Systems
     The research project is implemented by the International Institute for Environment and Development (IIED) with collaboration from the International Farming Systems Research Methodology Network (RIMISP) and the Royal Tropical Institute (KIT).
     The project analyses concentration in the processing and retail sectors of national and regional agrifood systems and its impacts on rural livelihoods. The role of primary producers and their economic organizations in negotiating market access and improving terms of trade in specific chains is examined.
     Because this programme looks into domestic and regional markets rather than

international markets, only a few abstracts have been included in the overview. Contact: http:/www.regoverningmarkets.org

Papers

17. T.A. Reardon, J.A. Berdegué, M. Lundy, P. Schütz, F. Balsevich, R. Hernández, E. Pérez and P. Jano. 2004. Briefing Paper: Supermarkets and Rural Livelihoods: A Research Method (as tested in Central America).

18. B. Vorley. 2004. Briefing Paper: Global dynamics of grocery retail restructuring: questions of governance.

19. D. Boselie and P. van der Kop. 2004. Briefing Paper: Institutional and organisational change in agri-food systems in developing and transitional countries: identifying opportunities for smallholders.

20. P. Rondot, E. Biénabe and M. Collion. 2004. Briefing Paper: Rural economic organizations and market restructuring: What challenges, what opportunities for small holders?

21. Regoverning markets. 2004. Final report of Phase 1
Contact: http://www.regoverningmarkets.org/docs/phase1.pdf

Regional papers and selected country studies
- East Africa, Kenya and Uganda
- Southern Africa, South Africa and Zambia
- Latin America, Guatemala - tomatoes, Nicaragua – tomatoes, Nicaragua -Cooperativa de Prodión y Comercialización, Guatemala - Asociación de Usuarios de Miniriego de Palencia
- China
- South Asia, Bangladesh and Pakistan
- South-East Asia, Philippines, Thailand and Viet Nam
- Central and Eastern Europe

Most of these reports are drafts for discussion. Only country studies which include the horticultural sector are listed here.
Contacts: http://www.regoverningmarkets.org/resources.html

Articles in journals

22. Berdegué, J.A. F. Balsevich, L. Flores and T. Reardon. 2005. Central American supermarkets' private standards of quality and safety in procurement of fresh fruits and vegetables. In. Food Policy. Vol. 30, nr.3, pages 254-269. Elsevier.

Keywords: Fresh produce; Supermarkets; Standards; Safety; Central America.

Abstract: Leading supermarket chains in Central America are imposing private standards. At the same time they are cutting costs through organizational change. The implementation of these private standards is good for consumers but a challenge for producers. Field study in Costa Rica, Guatemala, El Salvador, Honduras and Nicaragua from 2002 to 2004.

2. On methodology

Methodologies for quantifying impact of technical barriers to trade

4. Henson et al. 2002. Sanitary and Phytosanitary Requirements and Developing Country Agro-Food Exports: Methodological Guidelines for Country and Product Assessments. (see above under World Bank project).

23. Maskus, K.E. and J.S. Wilson (eds.). 2001. Quantifying the Impact of Technical Barriers to trade: Can it be Done? University of Michigan Press, Ann Arbor.

   Abstract: The emergence and enforcement of stronger standards among developed economies could result in diminished trade opportunities for developing countries. At the same time, standards could expand market access through resolving consumer information problems. Among the issues addressed in this collection are restrictions on genetically modified foods and pesticide use and compatibility standards for computers. The publication includes essays by prominent international trade specialists.

   23a. Maskus, K. E., J. S. Wilson and T. Otsuki (2001). An Empirical Framework for Analyzing Technical Regulations and Trade.

24. Beghin, J.C. and J-C. Bureau. 2001. Measurement of Sanitary, Phytosanitary and Technical barriers to Trade. OECD, Paris.

   Abstract: Assessment of various methodologies for measuring non-tariff trade barriers in the agricultural and food sectors: Analytical framework. The price-wedge method. Inventories-based approaches. Survey-based approaches. Gravity based approaches. Risk assessment-based cost-benefit measures. Stylised microeconomic approaches. Quantification using sectoral or multimarket models.

   Contact: Wayne.Jones@oecd.org - http://www.oecd.org/dataoecd/1/36/1816774.pdf

25 (=38). Henson, S.J. 2001. Measuring the Economic Impact of Technical Measures on Trade in Agricultural Commodities. FAO (ESC), Rome.

   Contact: http://siteresources.worldbank.org/INTRANETTRADE/Resources/Topics/Accession/Standards&TradeOverview_Eng.doc

26. Maskus, K. and J.S. Wilson. 2000. Quantifying the Impact of Technical Barriers to Trade: A Review of Past Attempts and the New Policy Context. World Bank, Washington DC.

27. Roberts, D., T. Josling, and D. Orden. 1999. A Framework for Analyzing Technical Trade Barriers in Agricultural Markets. Economic Research Services. United States Department of Agriculture, Washington, D.C.

Methodologies for value chain analysis

28. Barrientos, S., C. Dolan and A. Tallontire. 2003. A Gendered Value Chain Approach to Codes of Conduct in African Horticulture In. World Development Vol. 31, No. 9, pp. 1511–1526, 2003. Elsevier

   Key words: Africa, gender, codes of conduct, employment, export horticulture.

   Abstract: Combining global value chain and gendered economy approaches, this paper provides a framework, the "gender pyramid", for assessing the gender content of codes of conduct. It analyses labour codes currently applied in three industries exporting to Europe: South African fruit, Kenyan flowers and Zambian vegetables and flowers. It concludes that the gender sensitivity of the codes of conduct needs to be greatly enhanced.

Contact: Available at http:/www.sciencedirect.com

29. Kaplinsky, R. and M. Morris. 2001. A handbook for value chain research. Prepared for IDRC.

Abstract: What is a value chain and why is it important? Methodology: the point of entry for value chain analysis; mapping value; product segments; how producers access final markets; benchmarking production efficiency. Governance and upgrading in value chains: distributional issues, rents and barriers to entry.

Contact: Global value chain initiative hosted by the Institute for Development Studies (IDS): http://www.ids.ac.uk/globalvaluechains/index.html http://www.ids.ac.uk/ids/global/pdfs/VchNov01.pdf

Authors: kaplinsky@ids.ac.uk, morrism@ukzn.ac.za

30. Raikes, P., M. F. Larsen and S. Ponte. 2000. Global Commodity Chain Analysis and the French Filière Approach: Comparison and Critique. In. Economy and Society 29(3), 2000.

Abstract: Reviews two approaches to the study of economic restructuring which focus on commodity specific dynamics of change approach and the francophone "filière" tradition.

Contact: Centre for Development Research, Copenhagen, Denmark.

3. On the impact of governmental regulations

On the impact of import regulations: SPS measures

31. Henson, S. and J. S. Wilson (eds.). 2005. The WTO and Technical Barriers to Trade. 560 pp Edward Elgar Publishing.

Keywords: Standards, technical barriers, WTO, trade.

Selected articles:

PART I Theoretical And Quantification Issues

31a. Fischer, R. and P. Serra (2000), Standards and Protection

31b. (=24) Beghin J.C. and J-C. Bureau (2001), Quantitative Policy Analysis of Sanitary, Phytosanitary and Technical Barriers to Trade (see Chapter 3)

PART II Standards And Trade In Goods

31c. C. Perroni and R. Wigle (1999), International Process Standards and North-South Trade

PART III Sanitary And Phytosanitary Measures

31d. Wilson J.S. and T. Otsuki (2003), Food Safety and Trade: Winners and Losers in a Non-Harmonized World

31e. Henson S. and R. Loader (2001), Barriers to Agricultural Exports from Developing Countries: The Role of Sanitary and Phytosanitary Requirements

PART IV Institutional Issues

31f. Casella, A. (2001), Product Standards and International Trade. Harmonization through Private Coalitions?

31g. Vogel, D. (2001), Is There a Race to the Bottom? The Impact of Globalization on National Regulatory Policies

Contact: www.e-elgar.com (Ordered by ESCP)

32. Teisl, M. F. and J. A. Caswell. 2003. Information Policy and Genetically Modified Food: Weighing the Benefits and Costs. Working Paper 2003-01, Department of Resource Economics, University of Massachusetts, Amherst.

Keywords: GMOs, biotechnology, labelling, benefits, costs

Abstract: This paper discusses the merits of various studies of the costs of GMO labelling schemes, with a particular emphasis on the impact of the design of the labelling programme on benefits and costs. Mandatory versus voluntary. Positive versus negative. Segmentation based on testing or traceability (identity preserved) systems.

http://www.umass.edu/resec/workingpapers/Teisl%20&%20Caswell%20Working%20 Paper.pdf (or via http://www.umass.edu/resec/workingpapers/index.html)

33. Buzby, J. (ed.) 2003. International Trade and Food Safety: Economic Theory and Case Studies. United States Department of Agriculture. Agricultural Economic Report No. 828. Washington, D.C.

http://www.ers.usda.gov/publications/AER828/

Keywords: BSE, dioxin, economic theory, food safety, grains, international trade, meat, poultry, produce, regulation, regulatory trends, Salmonella, seafood, ERS, USDA. Selected chapters:

33a. Mitchell, L. Chapter 2. Economic Theory and Conceptual Relationships between Food Safety and International Trade
Abstract: The market generally does not provide socially desired levels of food safety. This gives reason for regulations, which impact on trade.
33b. Calvin, L. Chapter 5. Produce, Food Safety, and International Trade Response to U.S. Foodborne Illness Outbreaks Associated with Imported Produce.

34. Unnevehr, L. J. (ed.) 2003. Food Safety in Food Security and Food Trade. International Food Policy Research Institute, Washington, D.C.
http://www.ifpri.org/2020/focus/focus10.htm
Selected articles:
34a. Caswell. J.A. 2003 Trends in Food Safety Standards and Regulation: Implications for Developing Countries.
34b. Henson, S. Food Safety Issues in International Trade.
34c. Wilson, J.S. and T. Otsuki. 2003. Balancing Risk Reduction and Benefits from Trade in Setting Standards.
34d. Calvin, L., L. Flores, and W. Foster. 2003. Case Study: Guatemalan Raspberries and Cyclospora.
Abstract: Description of a public-private effort to solve a food-safety problem, where strict standards did not repair past reputation damage.
34e. Norton, G.W., G.E. Sanchez, D. Clarke-Harris and H. Koné Traoré. 2003. Case Study: Reducing Pesticide Residues on Horticultural Crops.
Abstract: three examples how to balance pesticide residue limits and phytosanitary requirements through IPM and pre-inspection protocols.
34f. Berdegué, J.A., F. Balsevich, L. Flores, D. Mainville and T. Reardon. 2003. Case Study: Supermarkets and Quality and Safety Standards for Produce in Latin America.
34g. Unnevehr, L.J., L. Haddad and C. Delgado. 2003. Food Safety Policy Issues for Developing Countries.

35. Josling, T., D. Roberts and D. Orden. 2003. Food Regulation and Trade: Toward a Safe and Open Global System. Institute for International Economics, Washington DC.

36. Wilson, J.S., T. Otsuki. 2002. To Spray or Not to Spray: Pesticides, Banana Exports, and Food Safety. Development Research Group (DECRG), World Bank.
Abstract: Gravity model results suggest that a 10 percent increase in regulatory stringency - tighter restrictions on the pesticide chlorpyrifos - leads to a decrease in banana imports by 14.8 percent. In addition, findings suggest that lack of consensus on international standards and divergent national regulations on pesticides is costly.
(Note: From a technical plant protection perspective a 15 percent import reduction is unlikely)

37. Otsuki, T., J.S. Wilson and M. Sewadeh. 2001. A Race to the Top? A Case Study of Food Safety Standards and African Exports. World Bank, Washington DC.

38 (=25). Henson, J. 2001. Measuring the economic impact of technical measures on trade in agricultural commodities, Working Paper, FAO (ESC), Rome.

39. Henson, S.J., R. Loader, A. Swinbank, M. Bredahl, and L. Lux. 2000. Impact of Sanitary and Phytosanitary Measures on Developing Countries. Department of Agricultural and Food Economics, University of Reading.

40. Otsuki, T., M. Sewadeh and J.S. Wilson. (2000) Saving Two in a Billion: A Case Study to Quantify the Trade Effect of European Food Safety Standards on African

Exports. World Bank, Washington DC.

Abstract: Uses the gravity equation method to determine the effects of European aflatoxin standards on African exports of dried fruits and nuts.

41. WTO. 2000. The Development Challenge in Trade: Sanitary and phytosanitary Standards. Submission by the World Bank. World Trade Organization, Geneva.

On the impact of import regulations: Organic and other standards

42. Garcia Martinez, M. and F. Bañados. 2004. Impact of EU organic product certification legislation on Chile Organic Exports. In. Food Policy, Vol. 29, No.1, pp.1-14

Keywords: Organic products; Chile; Certification legislation; Trade barriers; EU

Abstract: This paper presents the results of a study on the impact of EU organic certification legislation on Chilean organic exports. The lack of an equivalent system forces Chilean organic exports to enter the European Union through the "back door", that is, through special import permits, with the resulting increase in transaction costs as products accepted in one EU country may not be accepted in another. The paper reports also on the recent legislative developments to establish a national organic certification system in Chile and the problems encountered to make it operational.

Contact: marian.garcia@imperial.ac.uk

43. World Bank. 2001. Standards, Developing Countries and the Global Trading System. In. Global Economic Prospects and the Developing Countries 2001. World Bank, Washington DC

On the impacts of governmental regulations in export countries: labour rights

44. Kucera, D. and R. Sarna. 2004. How do trade union rights affect trade competitiveness? Working Paper No. 39. ILO

Abstract: The paper uses a bilateral trade gravity model to evaluate the effects of Freedom of Association and Collective Bargaining (FACB) rights and democracy on exports for the 1993 to 1999 period, including data for up to 162 countries. The paper finds robust relationships between stronger FACB rights and higher total manufacturing exports. However, the paper finds no robust relationship between FACB rights and labour-intensive manufacturing exports.

Contact:

http://www.ilo.org/public/english/bureau/integration/download/publicat/4_3_233_wp-39.pdf

45. Kucera, D. and R. Sarna. 2004. Child Labour, Education and Export Performance. Working Paper No. 52 ILO

Abstract: The paper uses a gravity trade model to estimate the effects of child labour and education on exports for the 1993 to 1999 period including data for up to 162 countries. This paper states that there is robust statistical evidence that child labour is bad and education is good for exports, including for unskilled labour-intensive manufacturing exports.

Contact: http://www.ilo.org/public/english/bureau/integration/download/publicat/4_3_302_wp-52.pdf

46. Chau, N.H. and R. Kanbur. 2000. The Race to the Bottom, From the Bottom

Abstract: South-South competition to export to the North, and its impact on labour conditions in developing countries. The paper argues that larger exporters will have better labour conditions for their workers than small exporters. (In the case that market requirements in the North do not include compliance with certain labour standards)

http://www.arts.cornell.edu/poverty/kanbur/ck15.pdf

4. On the impact of private and voluntary standards

47. Bazoche, P., E. Giraud-Héraud and L-G. Soler. 2005. Premium Private Labels, Supply Contracts, Market Segmentation, and Spot Prices, Journal of Agricultural & Food Industrial Organization: Vol. 3: No. 1, Article 7.

Abstract: European retailers have modified the market segmentation by implementing new private labels, imposing more demanding production requirements and relying on contractual relationships with upstream producers. This paper proposes a model of vertical relationships between producers and retailers in order to analyse the interest of producers to commit to these new private labels, their effects on spot market prices, and the resulting market segmentation between the spot market and supply contracts.

http://www.bepress.com/jafio/vol3/iss1/art7

48. Kilian, B., C. Jones, L. Pratt and A. Villalobos. 2005. The value chain for organic and fairtrade products and its implication on producers in Latin America. CIMS paper presented at the IAMA 15th Annual World Food & Agribusiness Symposium.

Abstract: Comparison of conventional, organic and fair-trade prices at FOB and retail levels for bananas and at farm gate, exporter, toaster and consumer level for coffee. Analysing the results with a standardized market model showed that price distortion along the value chain seriously affects the benefits distributed by sustainable production, favouring the retail and wholesale sectors instead of the production sector.

Contact: bernard.kilian@incae.edu, CIMS

http://www.ifama.org/conferences/2005Conference/Papers&Discussions/1042_Paper_Final.pdf

49. Basu, A., N. Chau and U. Grote. 2004. On export rivalry and the greening of agriculture: the role of eco-labels. Paper presented at the European Development Research Network (EUDN) second academic Conference on Trade, aid, FDI and international migration.

Keywords: Eco-labelling in Agriculture, Export Rivalry, Strategic Complementarity.

Abstract: (Note: the authors use the term "eco-labelling" to mean organic farming and other environmental certification programmes). The presented theoretical framework yields a set of empirical implications in a subgame perfect Nash equilibrium, and highlights: (i) the selection criteria of countries that adopt ecolabelling, and (ii) the endogeneity of labelling incentives and the welfare consequences of observed labelling initiatives. The theoretical findings are tested by investigating the time pattern of ecolabelling adoption by countries. Food industry export orientation appears to be correlated with the speed with which countries implement their own ecolabelling programmes.

Contact: http://www.eudnet.net/Member/afd_2004/Basu_Chau_and_Grote.pdf or http://www.eldis.org

50. Duprez, C. and J.M. Baland. 2004. Made in Dignity: The Effects of Labelling on Child Labour. European Development Research Network (EUDN)

Abstract: This paper analyses the impact of both social and geographical labelling on child labour. A simple model of North-South trade is developed, which shows that social labelling will not have any effects in several cases.

Contact: http://www.eudnet.net/Member/afd_2004/Duprez_and_Baland.pdf or http://www.eldis.org

51. Morrison, J., K. Cushing, Z. Day, and J. Speir. 2000. Managing a Better Environment: Opportunities and Obstacles for ISO 14001 in Public Policy and Commerce. Pacific Institute report. Oakland, California.
    Keywords: ISO 14001, trade, environment
    Abstract: Analysis of the creation of the standards, and their implications, benefits and limitations. The authors argue for changes in the ISO membership. Three case studies describe the emergence of ISO 14001 as an environmental regulatory tool in the United States. The authors make recommendations on how the ISO 14000 standards might be integrated into commercial practices, regulatory structures, and trade regimes in a socially equitable and environmentally beneficial manner.
    Contact: http://www.pacinst.org/topics/globalization_and_environment/public_policy/isoes.pdf

52. UNCTAD. 1997. Expert Meeting on Trade and Investment Impacts of Environmental Management Standards, particularly the ISO 14000 series, on Developing Countries. Geneva, 29-31 October 1997.
    52a. UNCTAD. 1997. Environmental management standards, Particularly the ISO 14000 series: Trade and investment impacts on developing countries
    Background Report prepared by the UNCTAD secretariat for the meeting.
    52b. UNCTAD. 1997. Report of the meeting TD/B/COM.1/10, TD/B/COM.1/EM.4/3
    http://www.unctad.org/en/docs/c1em4d3.en.pdf
    52c. UNCTAD. 1997. Recommendations adopted by the expert meeting.
    TD/B/COM.1/EM.4/L.1
    Contact: rene.vossenaar@unctad.org
    http://www.unctad.org/Templates/Meeting.asp?m=4220&intItemID=1942&lang=1

5. Case studies

Case studies on market entry barriers/reasons for adoption

53. Hattam, C. (PhD in progress) Small Farmer Organic Agriculture: Perceptions and Impacts of Certification.
   Keywords: Organic agriculture, small farmer, adoption
   Methodology: Household survey of organic and conventional producers of avocado in Mexico. Technology adoption theory, market entry work, the theory of planned behaviour for attitude variables.
   Contact: c.e.hattam@reading.ac.uk

54. Fouayzi, H., J.A. Caswell and N.H. Hooker. Forthcoming. Motivations of Fresh-Cut Produce Firms to Implement Quality Management Systems. In. Review of Agricultural Economics.

55. O'Brien, T.M. and A. Díaz Rodríguez. 2004. Mejorando la competitividad y el acceso a los mercados de exportaciones agrícolas por medio del desarrollo y aplicación de normas de inocuidad y calidad. El ejemplo del espárrago Peruano. Reporte del Programa de Sanidad Agropecuaria e Inocuidad de Alimentos del Instituto Interamericano de Cooperación para la Agricultura (IICA)
   Keywords: supply chain partnerships, quality standards, adoption, success case
   Contact: http://infoagro.net/shared/docs/a3/esparrago_peru.pdf

56 (=8). Henson, S. and S. Jaffee. 2004. Jamaica's Trade in Ethnic Foods and Other Niche Products: The Impact of Food Safety and Plant Health Standards.
   On SPS capacity as part of overall Jamaican competitiveness. (See WB project above)

57 (=9). Jaffee, S. 2003. From Challenge to Opportunity: Transforming Kenya's Fresh Vegetable Trade in the Context of Emerging Food Safety and other Standards in Europe
   On using standard compliance as a competitive advantage by Kenyan fresh produce industry. (See World Bank project above)

58 (=138). Borregaard, N., G. Geisse, A. Dufey and J. Ladron de Guevara. 2002. Green Markets. Often a Lost Opportunity for Developing Countries.
   Centre for Environmental Investigation and Planning (CIPMA)
   Keywords: organic; eco-labelling; wine; forest products; Chile; entry barriers; WTO
   Contact: IISD/ICTSD Trade Knowledge Network
   http://www.tradeknowledgenetwork.net/pdf/tkn_green_markets.pdf
   CIPMA: http://www.cipma.cl/

59. Kidd, A.D., A. Tulip and C. Walaga. 2001. Benefits of globalisation for poor farmers; A story of organic produce exports from Uganda. In. BeraterInnen News 2/2001: 25-31

60. López Figueroa, B. 2001. Estudio de Caso: Acción conjunta Ministerio de Agricultura de Guatemala y Asociaciones de Productores de Frambuesa, para Programas de Aseguramiento de Calidad Sanitaria e Inocuidad. Study for IICA.
   Abstract: Study of case to solve a food safety problem in raspberries from Guatemala.

Contact: http://infoagro.net/es/apps/casosexitosos/Frambuesas-Guatemala.doc

61. Julian, J.W., G.H. Sullivan and G.E. Sanchez. 2000. Future Market Development Issues Impacting Central America's Non-Traditional Agricultural Export Sector: Guatemala case Study. In. American Journal of Agricultural Economics, 82 (5), 1177-1183.

Cost/benefit analyses of standard implementation and certification

62. Anson, J., O. Cadot, A. Estevadeordal, J. de Melo, A. Suwa-Eisenmann, and B. Tumurchudur. 2005. Rules of Origin in North-South Preferential Trading Arrangements with an Application to NAFTA. In. Review of International Economics, 13 (3), pages: 501 - 517

Keywords: Rules of origin; cost of compliance; preferential trading agreements

Abstract: In the case of NAFTA, average compliance costs with rules of origin are found to be around 6 percent in ad-valorem equivalent, undoing the tariff preference (4 percent on average) for a large number of tariff lines.

Contact: Jose.anson@unil.ch; Olivier.Cadot@unil.ch; antonie@iadb.org;

demelo@ecopo.unige.ch; akiko.suwa@ens.fr; Bolormaa.TumurchudurKlok@unil.ch

6. Aloui, O., and L. Kenny. 2004. Case Study on Cost of Compliance to SPS Standards for Moroccan Exports.

On EurepGAP, HACCP, ISO9001 and BRC in Morocco tomato and citrus export sectors. (see World Bank project above)

11. Manarungsan et al. 2004. Costs of Compliance to SPS Standards: Thailand Case Studies of Shrimp, Fresh Asparagus, and Frozen Green Soybeans.

On the costs of meeting tighter Japanese rules on MRLs, and on the cost of converting to organic farming. (See World Bank project above)

12. Mbaye, A.A. 2004. Sanitary and Phytosanitary Requirements and Developing Country Agrifood Exports: An Assessment of the Senegalese Groundnut Sub-Sector.

On costs and benefits of meeting aflatoxin standards. (See World Bank project above)

63. Foli Gogoe, S. 2003. Costs and benefits of small-holders' compliance with the EurepGAP protocol in Ghana. MSc dissertation submitted to NRI, University of Greenwich, UK

Abstract: To comply with EurepGAP, growers faced high initial investment costs, resulting in higher fixed costs. Variable costs decreased. On average an 8 percent increase in profits was observed, but with a high variation between growers. The cost of training, certification and laboratory analysis were borne by the exporter. The acquired bookkeeping skills were highly appreciated.

Contact: Seth_Gogoe@sgs.com.

64. Damiani, O. 2001-2002. Series of case studies for IFAD of which the following involve fresh fruits & vegetables:

64a. Organic agriculture in El Salvador: the case of fresh vegetable in Las Pilas

Abstract: Technical assistance helped small farmers to convert a maize-vegetable system into a year-round organic vegetable production system for supermarkets in the capital. This led to higher labour demand and higher prices.

64b. Organic agriculture in Costa Rica: the case of cocoa and banana production in Talamanca.

Abstract: A farmer association was brought in contact with a buyer of organic cocoa. The buyer provided seed capital. This enabled the association to rehabilitate abandoned cocoa fields into mixed certified organic systems with bananas, fruits, tubers and shade trees.

64c. Small farmers and organic banana production in the Dominican Republic
Abstract: Two marketing firms contracted small-scale producer associations for which they managed organic certification. Compared with conventional small producers, organic producers faced on average 8 percent higher production costs but the price premium resulted in a 52 percent rise in the net revenue. However, the farmers had difficulties meeting increased quality demands and sometimes had to sell to the domestic market.
Contact: odamiani@usa.net

65. Collinson, C. 2001. The Business Costs of Ethical Supply Chain Management: Kenya Flower Industry Case Study Chatham, UK. NRI/NRET
Contact: http://www.nri.org/NRET/2607.pdf

66. Collinson, C. 2001. The Business Costs of Ethical Supply Chain Management: South African Wine Industry Case Study. Chatham, UK. NRI/NRET.
Contact: http://www.nri.org/NRET/2606.pdf

Case studies on impact versus objectives of the standards
67. Nelson, V., J. Ewert and A. Martin. 2002. Assessing the impact of codes of practice in the South African wine industry and Kenyan cut flower industry. Phase 1 report.
Chatham, UK. NRI/NRET
Contact: http://www.nri.org/NRET/phase1report.pdf

68. Nelson, V., A. Martin and J. Ewert. 2002. Methodological challenges to impact assessment of codes of practice. Paper presented at 5th Annual Warwick Corporate Citizenship Unit Corporate Citizenship Conference 2002.
Keywords: social impact, methodology, case studies
Contact: http://www.nri.org/NRET/methodological.pdf

69. Goldblatt, M., R. Cassim, M. Frimpong, M. Visser, P. Holden and A. Loewenther. 1999. Case study: The citrus exporting industry: an industry meeting international market. In. Trade and Industry Policy Secretariat – TIPS, IISD. 1999. Trade and Environment: South African Case-Studies. Page 43-48
Abstract: Discusses impacts of food quality and safety, environmental and social market requirements on the South African citrus industry, the domestic citrus market and domestic pesticide use.
Contact: http://www.tradeknowledgenetwork.net/pdf/sacasefullrprt_e.pdf

70. Blowfield, M. and Gallat, S. (around 1999). Volta River Estates Fairtrade Bananas case study. Ethical Trade and Sustainable Rural Livelihoods – case studies series. Chatham, UK. NRI/NRET
Contact: http://www.nri.org/NRET/csvrel.pdf

71. Malins, A. and M. Blowfield. (Around 1998). Fruits of the Nile, Fairtrade Processing case study. Ethical trade and sustainable rural livelihoods – case studies. Chatham, UK. NRI/NRET
Keywords: dried fruit, Uganda, fair-trade
Contact: http://www.nri.org/NRET/fruitnil.pdf

On food supply chains
Descriptive studies of supply chains and horticultural export industries
72. Cadilhon, J-J., A.P. Fearne, D.R. Hughes and P. Moustier. 2003. Wholesale Markets and Food Distribution in Europe: New Strategies for Old Functions. Discussion Paper No.2,

January 2003. Centre for Food Chain Research, Imperial College London.

Abstract: Wholesale markets in Europe (Paris, Rotterdam, London, Verona) are highly variable depending on: (i) the structure of final distribution, with variable importance of supermarkets, stores and restaurants. (ii) the ability of market managers to adapt to changes in demand (e.g. diversity and food safety). One-stop wholesale markets on the continent have not been losing as much market share to the cash-and-carry shops as the small product-specific sites in London. Furthermore, a proactive regulatory environment, the switch to high quality produce (including traceability) and a strong public relations policy has also sustained the sales of the Rungis wholesale market. http://www.imperial.ac.uk/agriculturalsciences/cfcr/pdfdoc/cadilhon2003.pdf

73. UNCTAD Diversification Programme.

Project implemented in 2000 and 2001. Descriptive papers of horticultural industries.

Contact: Mehmet Arda, mehmet.arda@unctad.org http://r0.unctad.org/infocomm/Diversification/

Workshop: Diversification et développement du secteur horticole en Afrique. Bamako, Mali, 13-15 February 2001.

73a. Simo, C., M. Yade and M. Sow. 2001. Perspectives des filières maraîchères au sahel: quelle dynamique de marché?

73b. D. Dembelé. 2001. Productions horticoles et perspectives de développement au Mali http://r0.unctad.org/infocomm/Diversification/Bamako/Dembele.PDF

Workshop: Diversification and development of the horticultural sector in Africa. Nairobi, Kenya, 29-31 May 2001

73c. Lheraut, G. 2001. Export logistics for ACP countries for fruit and vegetables and horticultural products. http://r0.unctad.org/infocomm/Diversification/nairobi/lherau.pdf

73d. Aloui, O. 2001. Performance in the agro-exports' sector: Tomatoes and strawberries in Morocco http://r0.unctad.org/infocomm/Diversification/nairobi/morocco.pdf

73e. Okado, M. 2001. Background paper on Kenya off-season and specialty fresh vegetables and fruits. http://r0.unctad.org/infocomm/Diversification/nairobi/keny0901.pdf

73f. Gritli, F. 2001. Horticulture in Tunisia: exports, incentives, financing instruments, marketing and support structures. http://r0.unctad.org/infocomm/Diversification/nairobi/gritli.pdf

73g. Schäfer, M. 2001. Organic production, processing and marketing.

73h. Keetch, D. P. 2001. The South African canned fruit and vegetable industry.

73i. Heri. S. T. 2001. The Growth and Development of the Horticultural Sector in Zimbabwe. http://r0.unctad.org/infocomm/Diversification/nairobi/horti_zimb.PDF

73j. Care International Kenya. 2001. The reap project: smallholder horticultural farming in Kenya.

Workshop: Commodity Export Diversification and Poverty Reduction in South and South-East Asia. Bangkok, Thailand, 3-5 April 2001.

73k. Hadi. P. U. 2001. The case study on canned pineapple in Indonesia.
http://r0.unctad.org/infocomm/Diversification/bangkok/pineap1.pdf

73l. Mathur. V. C. 2001. Export Potential of Onion: A Case Study of India.
http://r0.unctad.org/infocomm/Diversification/bangkok/onion.pdf

Workshop: El sector agroalimentario: Integración regional y vinculaciones internacionales para su desarrollo. Costa Rica 14-16 Marzo 2001.

73m. Loma-Ossorio Friend, E. de,. (PESA-FAO). 2001. La organización del sector agroalimentario como estrategia para el acceso a los mercados y la seguridad alimentaria en Centroamérica. http://r0.unctad.org/infocomm/Diversification/san%20jose/deloma.pdf

73n. Pomareda. C. 2001. Los Pequeños Productores Y Su Participación En Las Agroexportaciones En Centroamérica. http://r0.unctad.org/infocomm/Diversification/san%20jose/pomareda.pdf

73o. Gitli, E. and R. Arce. 2001. Consideraciones Sobre El Comercio Internacional De Los Productos Orgánicos En Centroamérica Ideas Sobre Costa Rica.

On retailer power in the food chain

74. Weldegebriel, H.T. 2004. Imperfect Price Transmission: Is Market Power really to Blame? In. Journal of Agricultural Economics Vol. 55, Nr. 1, March 2004, Pages 101-114 Agricultural Economics Society (winning entry of the 2002 Agricultural Economics Society Prize Essay)

Abstract: Develops a model of price transmission where both oligopoly and oligopsony power co-exist. It shows that taking the degree of price transmission in a perfectly competitive market as a benchmark, oligopoly and oligopsony power do not necessarily lead to imperfect price transmission, although they can.

Note: May be relevant to analyse distribution of certification costs through price transmission along the chain.

Contact: lexhab@nottingham.ac.uk PDF copy on file

75. DFID, T. Fox and B. Vorley. 2004. Concentration in food supply and retail chains. Working/discussion paper.

Keywords: buyer-driven chains; small producers; voluntary and regulatory standards

Contact: tom.fox@iied.org and bill.vorley@iied.org

76. Fearne, A., R. Duffy and S. Hornibrook. 2004. Measuring Distributive and Procedural Justice in Buyer/Supplier Relationships: An Empirical Study of UK Supermarket Supply Chains. Paper presented at the 88th Seminar of the European Association of Agricultural Economics. Retailing and Producer-Retailer Relationships in Food Chains Paris (France), May 5-6, 2004.

Keywords: Buyer/supplier relationships, trust, fairness, justice, UK supermarkets

Abstract: An empirical study of suppliers' perceptions of their trading relationships with the UK supermarkets. Conceptual framework based on Kumar's (1996) theory of justice. The results provide evidence of good practice in some supermarket relationships but show considerable room for improvement in others. Most significantly, the results suggest that a retail strategy based on low prices does not necessarily imply an abuse of market power or unfair treatment of suppliers.

Contact: a.fearne@imperial.ac.uk

http://www.racetothetop.org/documents/reports/Fearne_et_al_EAAE-Paris.pdf

See also briefing note: Methodology for quantitative comparison of UK multiple retailers' terms of trade with primary producers http://www.imperial.ac.uk/agriculturalsciences/cfcr/pdfdoc/brief3.doc

77. Fox T. and B. Vorley. 2004. Stakeholder accountability in the UK supermarket sector. Final report of the Race to the Top project

Race to the Top developed a benchmark for supermarkets with scoring methodology (indicators) on: environment, nature, consumer health, producers, workers (both within the company and in the supply chain) and local sourcing. Six UK chains participated in the pilot scoring, but only 3 in the second year, after which the project had to stop and tracking of progress was not possible.

Contact: http://www.racetothetop.org/documents/RTTT_final_report_full.pdf

78. Dolan, C. and J. Humphrey. 2001. Governance and Trade in Fresh Vegetables: The Impact of UK Supermarkets on the African Horticulture Industry. In. Journal of Development Studies 37(2)

79. Dolan, C., J. Humphrey and C. Harris-Pascal. Horticulture Commodity Chains: The Impact of the UK Market on the Fresh Vegetable Industry. IDS Working Paper 96
Abstract: In the United Kingdom, large supermarkets have captured most of the market for fresh vegetables. They specify cost, quality, delivery, product variety, innovation, and food safety and quality systems. The paper analyses how they have structured the horticulture export industries in Kenya and Zimbabwe.
Contact: http://www.ids.ac.uk/ids/bookshop/wp/wp96.pdf

80. Dolan, C., J. Humphrey. 2000 Changing Governance Patterns in the Trade in Fresh Vegetables between Africa and the United Kingdom
Abstract: Large UK retailers have adopted competitive strategies based on quality, year-round supply and product differentiation. Global value chain analysis is used to explain why supply chains have become much more vertically integrated. While the current trends may lead to a changing role for importers, the tendency towards the concentration of production and processing in Africa in the hands of a few large firms is likely to continue.
Contact: http://www.gapresearch.org/production/IFAMSubmission.pdf

81. UK Competition Commission. 2000. Supermarkets: A report on the supply of groceries from multiple stores in the UK.
Available at www.competition-commission.org.uk/reports/446super.htm. (See especially Chapter 11 and Appendix 11.3.)

On governance of food chains, including standards and certification
82. Ponte. S. (in press) Quality Conventions and the Governance of Global Value Chains
Keywords: global value chains, convention theory, governance, coordination, quality, standards, Africa
Abstract: Convention theory helps to understand governance in global value chains through analysis of "quality". "Lead" firms "drive" chains through relatively loose forms of coordination. They have been able to embed quality information into widely accepted standards, certifications, and codification procedures.
Contact: spo@diis.dk Danish Institute for International Studies
http://www.ids.ac.uk/globalvaluechains/publications/ponte-conventions.pdf

83. Ponte. S. (in press). Africa in the Age of Global Capitalism: Trade Rules, Value Chains and Quality Conventions. (Palgrave, forthcoming),

84. Humphrey, J. 2005. Shaping Value Chains for Development: Global Value Chains in Agribusiness. GTZ Trade programme. Eschborn.
Abstract: Challenges to reduce rural poverty in developing countries through increasing export of agricultural products arise in the areas of competition and from the increasing importance of standards in trade. Using a global value chain perspective, this study examines the implications of these challenges for policies (technical assistance, local institutional capabilities, producer organizations, etc.) and for the institutional framework that regulates agricultural production and trade, including standards-setting, intellectual property rights and global competition policy, as well as trade capacity building and trade promotion initiatives.
Contact: http://www2.gtz.de/dokumente/bib/05-0280.pdf

85. Hatanaka, M., C. Bain and L. Busch. 2005. Third-party certification in the global agrifood system. In. Food Policy. Vol. 30, nr.3, pages 354-369. Elsevier.
Keywords: Food safety; Standards; Certification
Abstract: Third Party Certification reflects the growing power of supermarkets to

regulate the global agrifood system. At the same time, it also offers opportunities to create alternative practices that are more socially and environmentally sustainable.

Contact: hatanaka@msu.edu, lbusch@msu.edu.

Corrected proof available at http://www.sciencedirect.com/

86. Sundkvist, A., R. Milestad and A. Jansson. 2005. On the importance of tightening feedback loops for sustainable development of food systems. In. Food Policy 30 (2005) 224–239. Elsevier

Keywords: Feedback; Food system; Management; Agriculture; Ecosystem; Society

Abstract: Discusses the importance of tightening feedback loops between ecosystems, actors in the food production chain and consumers. Where distances between resource and resource user are too large, feedback has to be directed through institutions on an overarching level, e.g., policy measures or environmental and social labelling of products.

Contact: asasun@infra.kth.se (Å. Sundkvist). Tel.: +46 8 79086 26.

87. Smith, G. C. and L. Saunders. 2005. International Identification, Traceability and Verification: The Key Drivers and the Impact On the Global Food Industry

Presented at the International Livestock Congress—2005 in Houston. Copyright: International Stockmen's Educational Foundation.

Contact: G.C. Smith, Center for Red Meat Safety, Colorado State University, Fort Collins, CO 80523-1171. L. Saunders, IMI Global, Inc., P.O. Box 1291, Platte City, MO 64079

www.livestockcongress.com

88. OECD. Programme on private standards and the agro-food system.

Work plan 2005-2006:

- Private standards and trade: consultants' report reviewing the use of private standards in sourcing agro-food products from developing countries, with a focus on Latin America

- Interplay between private standards and government regulation: stakeholder workshop followed by economic analysis of the various options for governance of the food sector.

88a. OECD. 2004. Private Standards And The Shaping Of The Agro-Food System. OECD Working Party on Agricultural Policies and Markets. AGR/CA/APM (2004)24

Keywords: private standards, food chain governance, policy issues

Abstract: Based on interviews with food retailers, standards' owners and manufacturers. Minimum quality standards set by government prompt higher quality firms to increase quality even further to remain competitive, but price premiums are reduced. However, food safety is seen by retailers as a non-competitive issue. Some private standards pre-empt regulation and often are later incorporated into regulation. According to interviewed GFSI members, they already require 100 percent certification from developing country suppliers.

Contact person: Linda Fulponi (E-mail: linda.fulponi@oecd.org)

89. Gereffi, G., J. Humphrey and T. Sturgeon. 2003. The Governance of Global Value Chains. In. Review of International Political Economy, Vol. 12, nr. 1, 2005, page 78.

Keywords: Global value chains; governance; networks; transaction costs; value chain modularity

Abstract: Theoretical framework that uses transaction costs economics, production networks, and technological capability and firm-level learning. Three important factors: (1) the complexity of transactions, (2) the ability to codify transactions, and (3) the capabilities in the supply-base. Five types of global value chain governance – hierarchy, captive, relational, modular, and market. Four case studies: apparel, bicycles,

horticulture and electronics.
   Contact: http://www.soc.duke.edu/~ggere/web/Governance_GVCs_RIPE_Feb%202005.pdf

90. Codron, J-M., E. Giraud-Heraud and L-G. Soler. 2003. French Large Scale Retailers and New Supply Segmentation Strategies for Fresh Products
   http://www.farmfoundation.org/documents/Jean-MarieCodron-final3-13-03_000.pdf

On labelling
On economics of labelling (economics of information)
91. Grolleau, G. and J. A. Caswell. 2005. Interaction Between Food Attributes in Markets: The Case of Environmental Labeling. University of Massachusetts Amherst,
Department of Resource Economics, Working Paper No. 2005-5
   Keywords: Environmental labelling, food attributes, food marketing, quality perception
   Abstract: Results suggest that the market success of eco-friendly food products requires a mix of environmental and other verifiable attributes that together signal credibility.
   Contact: g.grolleau@enesad.inra.fr tel. (33)380-772443,
   caswell@resecon.umass.edu tel. (1)413-545-5735
   http://www.umass.edu/resec/workingpapers

92. Grolleau, G. and J. A. Caswell. 2003. Giving Credence to Environmental Labeling of Agro-Food Products: Using Search and Experience Attributes as an Imperfect Indicator of Credibility. In. Ecolabels and the Greening of the Food Market, ed. W. Lockheretz, pp. 121-129. Boston, MA: Tufts School of Nutrition Science and Policy.
   http://nutrition.tufts.edu/pdf/conferences/ecolabels/proceedings.pdf

93. Bonroy, O. and M. Laclau. 2002. Quality and label : The Case of Credence Goods. CATT, University of Pau
   Keywords: Asymmetric information, credence goods, consumers' beliefs, label.
   Contact: olivier.bonroy@univ-pau.fr marc.laclau@univ-pau.fr

94. Krissoff, B., M. Bohman and J.A. Caswell (eds.). 2002. Global Food Trade and Consumer Demand for Quality. Kluwer Academic, New York (see also 119)

95. Bramley-Harker, E., J. Dodgson and M. Spackman. 2001. Economic Appraisal of Options for Extension of Legislation on GM Labelling. Report by National Economic Research Associates for the UK Food Standards Agency.
   Abstract: Multicriteria analysis of costs (including enforcement costs and distributional effects along the chain), benefits, risks and uncertainties associated with a number of options for GM labelling. Options evaluated: status quo, voluntary GM-free, mandatory GM (derived from/derived with help from).
   Contact: Edward.bramley.harker@nera.com, john.dodgson@nera.com,
   michael.spackman@nera.com; http://www.foodsafetynetwork.ca/gmo/fsagmlbl.pdf;

96. Caswell, J. A. 2000. Analyzing Quality and Quality Assurance (Including Labeling) for GMOs. AgBioForum. Published by Illinois Missouri Biotechnology Alliance 3 (4/Winter).
   Key words: GMOs; quality assurance; labelling.
   Contact: http://www.agbioforum.org/v3n4/v3n4a08-caswell.htm
   (413)545-5735, caswell@resecon.umass.edu
   http://www.umass.edu/resec/faculty/caswell/#Economics%20of%20Food%20Labeling

On markets for labelled products

97. Willer, H. and M. Yussefi (Eds.). 2005. The World of Organic Agriculture Statistics and Emerging Trends 2005. IFOAM/FiBL
    Yearly publication.
    Contact: Hard copies can be ordered online or the full document can be downloaded at: www.ifoam.org or www.fibl.org/english/shop/

98. Centro de Inteligencia de Mercados Sostenibles (CIMS)
    CIMS publishes regular market studies for organic and other certified products which are of interest to Latin America. These studies are for sale, but some can be downloaded for free when registered (registration is for free). All documents are available in Spanish, some also in English. Recent studies:
    98a. Perfil de mercado de piña sostenible 2005
    98b. Análisis del mercado del aguacate convencional y orgánico en la Unión Europea
    98c. Requisitos y regulaciones para la importación de frutas tropicales a los Estados Unidos: Capítulos específicos para: aguacate, banano, mango, papaya
    98d. Análisis del mercado de papaya convencional y orgánica en la Unión Europea
    Contact: http://www.CIMS-LA.com

99. Rozan, A., A. Stenger and M. Willinger. 2004. Willingness-to-pay for food safety: an experimental investigation of quality certification effects on bidding behaviour. In. European Review of Agricultural Economics. Vol. 31(4) Wageningen.
    Contact: http://www.sls.wau.nl/aae/erae/erae_issues.htm

100. Lusk, J.L., L.O. House, C. Valli, S.R. Jaeger, M. Moore, B. Morrow W. and B. Traill. 2004. Heterogeneity in Consumer Preferences as Impetus for Non Tariff Trade Barriers: Experimental Evidence of Demand for Genetically Modified Food in the United States and European Union
    Keywords: GM food, experimental auction
    Abstract: The median level of compensation demanded by English and French consumers to consume a genetically modified food was more than twice that in any of the United States locations.
    Contact: jlusk@purdue.edu http://www.agecon.purdue.edu/staff/jlusk/USEU%20AJAE.pdf

101. Umberger, W.J. and Feuz D.M. 2004. The Usefulness of Experimental Auctions in Determining Consumers' Willingness-to-Pay for Quality-Differentiated Products. In. Review of Agricultural Economics, Vol.26 nr.2 page 170.
    Abstract: The validity and effectiveness of using experimental auctions to elicit consumers' willingness-to-pay for closely related, quality-differentiated products is examined. Demographic variables are poor predictors of bids and auction winners. Panel size and initial endowment influence auction results. Relative willingness-to-pay values elicited through experimental auctions appear valid, while actual willingness-to-pay values are influenced by experimental design.
    Contact: http://www.blackwell-synergy.com/toc/raec/26/2

102. Consumers International. 2004. Green Food Claims. An international survey of self-declared green claims on selected food products. Consumers International's Office for Developed and Transitional Economies (ODTE).
    Abstract: The paper explores whether European and US consumers can trust the information displayed on everyday food labels/packaging.
    Contact: www.consumersinternational.org

103. Sligh, M. and C. Christman. 2003. Who Owns Organic? The Global Status, Prospects, and Challenges of a Changing Organic Market. RAFI-USA
   Keywords: Organic market, US, concentration
   Abstract: Includes an overview of the corporate structure of the bigger organic food companies and retailers and concentration trends.
   Contact: http://www.rafiusa.org/pubs/OrganicReport.pdf,
   Michael Sligh, msligh@rafiusa.org

104. UNCTAD. 2003. Organic Fruit and Vegetables from the Tropics. Market, Certification and Production Information for Producers and International Trading Companies.
   Abstract: Organic production practices by crop. Brief market outlook and certification requirements.

105. Lohr, L. 2001. Factors Affecting International Demand and Trade in Organic Food Products. In. Changing Structure of Global Food Consumption and Trade. A. Regmi (ed.). ERS WRS No. 01-1, USDA Economic Research Service, Washington, DC, p. 67-79.

106. Lohr, L. 2001. The Importance of the Conservation Security Act to U.S. Competitiveness in Global Organic Markets. FS 01-19, Dept. of Agricultural and Applied Economics, University of Georgia.

107. Lohr, L. 2001. Predicting Organic Market Development with Spatial Analysis of Existing Industry Information. FS 01-15, Dept. of Agricultural and Applied Economics, University of Georgia.

On standard setting and the design og conformity assessment programmes
108. Courville, S. (forthcoming). Standards and Certification. In. Kristiansen, Paul and Acram Taji (eds.) Organic Agriculture: A Global Perspective. Collingwood: CSIRO Publishing.

109. Proforest. 2005. Managing conflict of interest in certification. A report for the ISEAL Alliance.
   http://www.isealalliance.org/documents/pdf/COI_Feb05_PD1.pdf

110. Pi Environmental Consulting. 2004. Learning from Social and Environmental Schemes for the ECL Space: Knowledge Base synthesis report.
   Keywords: environmental and social requirements, access, scheme typology
   Abstract: Review of 95 existing case studies on impact and market access of environmental and social requirements. Analysis of the relation between typology of the certification scheme (mandatory/voluntary, stakeholder participation, transparency etc.) and the extent to which it acts as barrier to trade.
   Background case studies:
   110a. Amariei, L. 2004. Learning from Social and Environmental Schemes for the ECL Space: ETI and EurepGAP case studies.
   110b. Quinoñes, B.R. 2004. Learning from Social and Environmental Schemes for the ECL Space: FLO case study.
   110c. Raste, A. 2004. Learning from Social and Environmental Schemes for the ECL Space: IFOAM case study.
   110d. Acuña, E. 2004. Learning from Social and Environmental Schemes for the ECL Space: ISO 14001 case study.
   Contact: Pierre Hauselmann phauselm@piec.org
   http://www.piec.org/ecl_space/07-CG_section/Knowledgebase/knowledgebase.pdf

http://www.piec.org/ecl_space/07-CG_section/casestudies.html

111. Michaud, J., E. Wynen and D. Bowen (eds.). 2004. Harmonization and Equivalence In Organic Agriculture - Volume 1. FAO, IFOAM and UNCTAD.
Selected contributions:
111a Courville, S. and D. Crucefix (2004). Existing and Potential Models and Mechanisms for Harmonization, Equivalency and Mutual Recognition.

112. Courville, S. 2004. Making Sense of Corporate Responsibility Tools. In. Galea, Chris (ed.) Teaching Business Sustainability –Volume 1. Sheffield: Greenleaf.

113. Smith, G. and D. Feldman. 2004. Implementation mechanisms for codes of conduct. Study prepared for the CSR Practice, Foreign Investment Advisory Service
Investment Climate Department, The World Bank/International Finance Corporation
Keywords: CSR; monitoring; apparel; footwear; agribusiness
Abstract: With the exception of a few leading initiatives, monitoring of the implementation of codes of conduct in the agribusiness sector is close to nonexistent. Unique to this sector is the use of large external NGOs as third-party verifiers. Common problems (all industries):
  • A lack of convergence in codes and in the training of monitors
  • The need for more multi-stakeholder initiatives
  • Disagreements over how far down the supply chain companies should monitor
  • Insufficient transparency and inadequate education of workers on their rights
Contact: http://www.ifc.org/ifcext/economics.nsf/AttachmentsByTitle/Implementation+mechanisms/$FILE/Implementation+mechanisms.pdf

114. Jahn, G., M. Schramm and A. Spiller. 2004. Trust in Certification procedures: An Institutional Economics Approach Investigating the Quality of Audits within Food Chains 2004. Paper presented at the IAMA World Food & Agribusiness Symposium 2004
Keywords: Certification, Audit Theory, Institutional Economics, Low Balling-Effect
Abstract: Only a reliable control procedure can reduce the risk of food scandals. This paper presents a model to enhance the efficiency of certification systems building on findings from financial auditing and the theory of the New Institutional Economics. Dumping prices on the certification market and differences in performance reveal the need for changes. Strategies are suggested to reduce auditors' dependence, intensify liability, increase reputation effects and minimise audit costs.
Contact: a.spiller@agr.uni-goettingen.de
http://www.ifama.org/conferences/2004Conference/Papers/Spiller1023.pdf

115. Jahn, G., M. Schramm and A. Spiller. 2004. Differentiation of Certification Standards: The trade-off between generality and effectiveness in certification systems. Paper presented at the IAMA World Food & Agribusiness Symposium 2004
Keywords: Certification, Information Economics, Crowding Effect, Harmonization
Abstract: A growing number of certification systems indicates the importance of third party audits but implies the danger of "audit tourism" and, as a consequence, rising transaction costs. The driving forces of this differentiation process are analysed. The trade-off between generality of a system and its effectiveness is revealed, which can be traced back to the disadvantages of general management system audits.
Contact: gjahn@gwdg.de

http://www.ifama.org/conferences/2004Conference/Papers/Jahn1024.pdf

116. ISEAL Alliance. 2003. Setting Social and Environmental Standards: A Research Report on Existing Standard-setting Practices.
http://www.isealalliance.org/documents/pdf/R028_PD1.pdf

117. Barling, D. and T. Lang. 2003 Codex, the European Union and Developing Countries: an analysis of developments in international food standards setting. Department of Health Management and Food Policy Institute of Health Sciences, City University.
A report for the Programme of Advisory Support Services (PASS) for the Rural Livelihoods Department of the UK Department for International Development (DfID). PASS project code TR0033.
Keywords: Codex, food safety, barriers.
Contact: d.barling@city.ac.uk, t.lang@city.ac.uk

118. Courville, S. 2003. Social Accountability Audits: Challenging or Defending Democratic Governance? In. Law and Policy 25(3). Pp. 267-297.

119. Lohr, L. and B. Krissoff. 2001. Consumer Welfare Effects of Harmonizing International Standards for Trade in Organic Foods. In. Global Food Trade and Consumer Demand for Quality. B. Krissoff, M. Bohman and J.A. Caswell (eds.). Kluwer Press (= 94.)
Contact: llohr@agecon.uga.edu http://www.agecon.uga.edu/faculty/llohr/index.html

120. Barrientos, S., C. Dolan and A. Tallontire. 2001. Gender and Ethical Trade: A Mapping of the Issues in African Horticulture. Chatham, UK. NRI/NRET
Keywords: South Africa, Kenya, Zambia, export, gender, codes.
Contact: http://www.nri.org/NRET/genderet.pdf

On relations between private standards and (inter)governmental standards
On relations between private standards and governmental regulations
121. Codron, J-M., E. Giraud-Héraud and L-G. Soler. 2005. Minimum quality standards, premium private labels, and European meat and fresh produce retailing. In. Food Policy. Vol.30, nr.3, pages 270-283. Elsevier.
Keywords: Food quality; Standards; Labelling
Abstract: The nature and determinants of retailer strategies following the mad cow crisis compared with food safety strategies in the fresh produce sector. Analysis of how the levels and enforcement of governmental standards influence the strategies of retail chains. Conclusions regarding what variables governments should take into account when they define minimum quality standards.
Contact: codron@ensam.inra.fr, giraude@poly.polytechnique.fr, soler@ivry.inra.fr.
Corrected proof available at http://www.sciencedirect.com/

122. Mainville, D.Y., D. Zylbersztajn, E.M.M.Q. Farina and T. Reardon. 2005. Determinants of retailers' decisions to use public or private grades and standards: Evidence from the fresh produce market of São Paulo, Brazil. In. Food Policy, Vol.30, nr.3, pages 334-353.Elsevier.
Keywords: Retailers; Food; Grades and standards; Fresh produce; Brazil
Abstract: The importance of the product in the firm's activities or sales, market power, scale of operations and investment in brand capital and reputation are key firm-specific factors encouraging the use of private G&S regimes over public.

Contact: mainvill@vt.edu, reardon@msu.edu
Corrected proof available at http://www.sciencedirect.com/

123. Courville, S. (forthcoming). Understanding NGO-Based Social and Environmental Regulatory Systems: Why We Need New Models of Accountability. In: Dowdle, C. (ed.) Rethinking Public Accountability. Cambridge: Cambridge University Press.

124. Fetter, T.R. and J.A. Caswell. 2002. Variation in Organic Standards Prior to the National Organic Program. In. American Journal of Alternative Agriculture 17(2):55-74.

125. Mojduszka, E.M. and J.A. Caswell. 2000. A Test of Nutritional Quality Signaling in Food Markets Prior to Implementation of Mandatory Labelling. In. American Journal of Agricultural Economics 82(May):298-309.

On relations between private standards and international agreements
126. Vallejo, N., J. Morrison and P. Hauselmann. 2004. Certification and Trade Policy Strategic Assessment. Report by Pi Environmental Consulting and the pacific Institute for the ISEAL Alliance. Public version edited by ISEAL.
     Contact: www.isealalliance.org nvallejo@piec.org jmorrison@pacinst.org phauselm@piec.org2004

127. UNCTAD. 2004. Environmental Requirements and Market Access for Developing Countries. Note by the UNCTAD secretariat for the pre-UNCTAD XI workshop on Environmental Requirements and Market Access for Developing Countries. TD/(XI)/BP/1
     Abstract: Environmental requirements are becoming more frequent, stringent and complex. Governmental standards and regulations, which fall under the TBT Agreement, represent only a small part of environmental requirements. There is a need to (a) more effectively involve developing countries in standard setting; (b) improve access to information on environmental requirements; and (c) strengthen developing countries' response capacities by promoting proactive adjustment policies.
     Contact: http://www.unctad.org/en/docs/tdxibpd1_en.pdf

128. Roberts, D. 2004. The Multilateral Governance Framework for Sanitary and Phytosanitary Regulations: Challenges and Prospects. World Bank, Washington DC.

129. Rotherham. T. 2003. Labelling for Environmental Purposes: A review of the state of the debate in the World Trade Organization - Full Report
     Contact: http://www.tradeknowledgenetwork.net/pdf/tkn_labelling.pdf

130. IATRC (2001). Agriculture in the WTO: The Role of Product Attributes in the Agricultural Negotiations. International Agricultural Trade Research Consortium, St Paul.

On policy options
131. Henson, S., and T. Reardon. 2005. Private agri-food standards: Implications for food policy and the agri-food system. In. Food Policy, Vol. 30, nr.3, pages 241-253. Elsevier.
Keywords: Private standards; Food safety; Food quality; Trade
     Abstract: Introduction to the evolution and nature of private food safety and quality standards, highlighting the resultant impacts on the structure and modus operandi of supply chains. Introduction to a series of papers.
     Contact: shenson@uoguelph.ca, reardon@msu.edu.
     Corrected proof available at http://www.sciencedirect.com/

132. Giovanni, D. and S. Ponte. 2005. Standards as a new form of social contract? Sustainability initiatives in the coffee industry. In. Food Policy, Vol. 30, nr.3, pages 284-301. Elsevier.

   Keywords: Grades and standards; Coffee

   Abstract: In the past, markets were embedded in a normative framework generated by government and labour unions. Now, standard-setting processes operate as new forms of social contract. This case study addresses standards' effectiveness in creating new markets, addressing collective and private interests and delivering sustainability. What is the role of public policy?

   Contact: dgiovanni@worldbank.org, spo@tele2adsl.dk (S. Ponte).

   Corrected proof available at http://www.sciencedirect.com/

133. Garcia Martinez M. and N. Poole. 2004. The development of private fresh produce safety standards: Implications for developing Mediterranean exporting countries. In. Food Policy, Vol. 29, No. 3, pages 229-255

   Keywords: Private standards; Food safety; Developing Mediterranean countries; Fresh produce

   Abstract: Part of EU research project. This article examines the impact of increasing demands for food safety and quality by European food retailers. The fundamental structure and culture of supplier organizations required by European retail chains are major entry barriers for developing Mediterranean fresh produce exporting countries and for developing countries in general. To sustain international demand for their products they have to take structural, strategic and procedural initiatives.

   Contact: marian.garcia@imperial.ac.uk, available online at http://www.sciencedirect.com

134. Henson, S. and M. Bredahl. 2004. Policy Options for Open Borders in Relation to Animal and Plant Protection and Food Safety. In. Loyns, R.M.A., Meilke, K., Knutson, R.D. and Yunez-Naude, A. (eds.). Keeping the Borders Open. University of Guelph, Canada.

135. UNCTAD. 2004. Trading opportunities for organic food products from developing countries. Strengthening research and policy-making capacity on trade and environment in developing countries.

   Abstract: There are important potential benefits of organic agriculture in developing countries. For export development, however, complex import and certification / accreditation procedures need to be addressed. Subsidies for organic agriculture in developed countries impact competitiveness. Comprehensive policies at both national and international levels are required. For small developing countries and LDCs, assistance from donors, as well as sharing the costs of certification with developed country partners (e.g. projects, fair-trade), may be the preferred option. For developing countries with a relatively large organic potential, developing a national certification system may be a priority.

   Contact: http://r0.unctad.org/trade_env/test1/publications/organic.pdf?docid=450 2%2B%22ItemID=2068)

136. Wilson, J.S. and V.O. Abiola. 2003. Standards and Global Trade: A Voice for Africa. World Bank, Washington DC. (See Chapter 2 under World Bank)

137. Pick, D. 2003. Product Differentiation and Asymmetric Information in Agricultural and Food Markets: Defining the Role for Government: Discussion In. American Journal of Agricultural Economics Vol. 85 Issue 3 Page 742 August 2003

138. (=58.) Borregaard et al. 2002. Green Markets. Often a Lost Opportunity for

Developing Countries. (see Case studies on market entry barriers, Chapter 6)

139. UNCTAD. 2001. Expert Meeting on Ways to Enhance the Production and Export Capacities of Developing Countries of Agriculture and Food Products, Including Niche Products, such as Environmentally Preferable Products. Geneva, July 2001
    139a. UNCTAD. 2001. Expert Meeting Report: TD/B/COM.1/41, TD/B/COM.1/EM.15/3.
    Abstract: Summary of conclusions and recommendations: policy options for Governments, the international community and UNCTAD. http://www.unctad.org/en/docs/c1em15d3.en.pdf
    139b. UNCTAD. 2001. Expert Meeting Background note by the Secretariat.
    Abstract: World markets are increasingly competitive and demanding, with a multitude of standards to be met. New skills are required in both the public and the private sectors and this often requires international cooperation. Rapid demand growth for organic food is likely to create temporary supply shortages. Such short-term opportunities can, however, only be seized if certification requirements can be met. As long as developing country producers retain significant production cost advantages, they might be able to consolidate the market shares gained in the short term.

140. Caswell, J.A. 2000. Labeling Policy for GMOs: To Each His Own? In. AgBioForum Vol. 3 (1) 53-57.
    Key words: GMOs; biotechnology; labelling policy; trade disputes.
    http://www.agbioforum.org/v3n1/v3n1a08-caswell.htm

On technical assistance
141. Wiig, A, and I. Kolstad. 2005. Lowering barriers to agricultural exports through technical assistance. In. Food Policy 30 (2005) 185–204. Elsevier
    Keywords: Food policy; Agriculture; Exports; Aid evaluation
    Abstract: Under the SPS Agreement, developed countries are to provide technical assistance to developing countries, to help them meet SPS requirements. A survey reveals, however, that assistance is allocated in an ad hoc manner. Data is presented which highlights the major problems of developing countries in exporting to the European Union and the United States.
    Contact: arne.wiig@cmi.no, ivar.kolstad@cmi.no

142. MSU, Institute of International Agriculture. EurepGAP Certification Study
Abstract: USAID-funded study under the RAISE/SPS IQC on the effects of third party certification on developing countries. The goal is to enhance USAID's capacity to assist smallholders, agribusinesses and government agencies to seed in meeting the challenges of private standards imposed by the supermarket sector.
Interviews being conducted August 2005
Contact: Dr Deepa Thiagarajan at thiagara@msu.edu Tel. (1) 517 432 8211

143. ITC and Commonwealth Secretariat. 2003/2004. Influencing and Meeting International Standards: Challenges for Developing Countries. Vol. 1, 2003: Background Information, Findings from Case Studies and Technical Assistance Needs. Vol. 2, 2004: Procedures Followed by Selected International Standard-Setting Organizations and Country Reports on TBT and SPS. International Trade Centre, Geneva
    Abstract: Conclusions and recommendations for technical assistance based on case studies on standards and quality management conducted in Jamaica, Kenya, Malaysia, Mauritius, Namibia and Uganda.
    Contact: http://www.intracen.org/eshop/f_e_Publications.asp?LN=EN
    http://publications.thecommonwealth.org/publications/html/DynaLink/cat_id/44/pub_id/341/pub_details.asp

144. Rotherham, T. 2003. Implementing Environmental, Health and Safety (EH&S) Standards, and Technical Regulations: The Developing Country Experience
    Keywords: Standards; Conformity assessment; Institutional infrastructure; WTO; harmonization; technical assistance
    Contact: http://www.tradeknowledgenetwork.net/pdf/tkn_standards.pdf

145. Gibbon. P. 2003. Commodities, donors, value-chain analysis and upgrading.
    Paper prepared for UNCTAD
    Abstract: There is a need for national sector-wide organization and assistance. This implies in-country and international donor coordination. The recent emphasis on encouraging production for niche markets has been probably excessive. While support to large cooperatives and large outgrower schemes offer possibilities to be more socio-economically inclusive, it should be recognised that they embody dynamics of differentiation and therefore internal marginalization. More attention should be given to complementary international interventions. These could range from consumer market development and removal of subsidies in the North to unbundling global market oligopolies.
    Contact: pgi@diis.dk

64. Damiani O. 2001-2002. Series of case studies for IFAD
    In addition to a cost benefit analysis these case studies also analyse the existence of supporting institutions and the types of technical assistance that has been received by the concerned farmers. (See Case studies on cost benefit analysis)

Other literature
146. IAAS. 2003. Food Quality a Challenge for North And South. Proceedings of the IAAS World Congress 2003. 319 pages.
    Part I: Food quality and agriculture
    146a. Schnug, E. 2003. Organically grown crops in the South – challenges and implications. page 81. Part II: Food quality and food industry
    146b. Gellynck, X., W. Verbeke and J. Viaene. 2003. Interactions in the food chain: towards integrated quality management. page 121
    146c. Mathijs, E. 2003. Marketing food quality: the role of labels and short chains. page 157
    Part III: Food quality and the consumer
    Part IV: food quality and food policy
    146d. Kabwit Nguz, A. 2003. Food safety and international trade: a challenge for developing countries. page 293
    Contact: IAAS Belgium vzw info@iaas.be
    Order address: Kasteelpark Arenberg 20, B-3001 Heverlee, Belgium

147. Commission on Human Rights. 2005. Report of the United Nations High Commissioner on Human Rights on the responsibilities of transnational corporations and related business enterprises with regard to human rights. CHR 61st session, Item 16 of the provisional agenda. E/CN.4/2005/91
    Keywords: Business; human rights; responsibility; standards

Abstract: The report reviews existing initiatives and standards on corporate social responsibility from a human rights perspective, noting that there are gaps in understanding the nature and scope of the human rights responsibilities of business.
Contact: http://www.ohchr.org/english/bodies/chr/sessions/61/lisdocs.htm

http://daccessdds.un.org/doc/UNDOC/GEN/G05/110/27/PDF/G0511027. pdf?OpenElement

148. Fearne A, M. Garcia Martinez, N. Bourlakis, M. Brennan, M. Temple and L. de Motte 2004. Mapping of Potential Research Areas for Economics Analysis. Document prepared for the Food Standards Agency under the contract RRD10/D03/B. Imperial College London, 16 March 2004
Contact: a.fearne@imperial.ac.uk

http://www.food.gov.uk/science/research/researchinfo/supportingresearch/ economics/economicresearch/d03projdetails/d03003/ (report not available on the web)

149. Fearne A, M. Garcia Martinez, N. Bourlakis, M. Brennan, J.A. Caswell, N. Hooker and S. Henson. 2004. Review of the Economics of Food Safety and Food Standards. Document prepared for the Food Standards Agency under the contract RRD10/D03/A. Imperial College London, 24 February 2004
Contact: a.fearne@imperial.ac.uk

http://www.food.gov.uk/science/research/researchinfo/supportingresearch/ economics/economicresearch/d03projdetails/d03002/ (report not available on the web)

Annex 1 - Alphabetical list of referenced authors

The authors of the studies quoted in the overview above are listed below in alphabetical order. In the overview, the studies have been numbered. If the author was first author of a publication, the number of the publication appears in the second column. If the author was co-author of a publication, the number of the publication appears in the third column (other reference). (Note: an author may have published on other subjects that were not listed in the overview and thus do not appear in this table).

Author	First author reference	Other references
Abiola, V.O.		5 (=136)
Acuña, E.	110d	
Aloui, O.	6, 73d	
Amariei, L.	110a	
Anson, J.	62	
Arce, R.	73o	
Bain, C.		85
Baland, J.M.		50
Balsevich, F.		17, 22, 34f
Bañados, F.		42
Barclay, R.		13
Barling, D.	117	
Barrientos, S.	28, 120	
Basu, A.	49	
Bazoche, P.E.	47	
Beghin, J.C.	24 (=31b)	
Berdegué, J.A.	22, 34f	17
Biénabe, E.		20
Blowfield, M.	70	71
Bohman, M.		94
Bonroy, O.	93	
Borregaard, N.	58 (=138)	
Boselie, D.	19	
Bourlakis, N.		148, 149
Bowen, D.		111
Bramley-Harker, E.	95	
Bredahl, M.		39, 134
Brennan, M.		148, 149
Bureau, J-C.	15	24 (=31b)
Busch, L.		85
Buzby, J.	33	
Cadilhon, J-J.	72	
Cadot, O.		62
Calvin, L.	33b, 34d	
Canale, F.	7	
Care International Kenya	73j	
Casella, A.	31f	
Cassim, R.		69
Caswell, J.A.	34a, 96, 140	32, 54, 91, 92, 94, 124, 125, 149
Chau, N.H.	46	49
Christman, C.		103
CIMS	98	
Clarke-Harris, D.		34e
Codron, J-M.	90, 121	
Collinson, C.	65, 66	
Collion, M.		20
Commission on Human Rights	147	

Author	First author reference	Other references
Commonwealth Secretariat		143
Consumers International	102	
Courville, S.	108, 111a, 112, 118, 123	
Crucefix, D.		111a
Cushing, K.		51
Damiani, O.	64	
Day, Z.		51
Delgado, C.		34g
Dembelé, D.	73b	
DfID	75	
Díaz Rodrígues, A.		55
Dodgson, J.		95
Dolan, C.	78, 79, 80	28, 120
Dufey, A.		58
Duffy, R.		76
Duprez, C.	50	
Estevadeordal, A.		62
Ewert, J.		67, 68
Farina, E.M.M.Q.		122
Fearne, A.P.	76, 148, 149	72
Feldman, D.		113
Ferner, V.		3
Fetter, T.R.	124	
Feuz, D.M.		101
Fischer, R.	31a	
Flores, L.		22, 34d, 34f
Foli Gogoe, S.	63	
Foster, W.		34d
Fouayzi, H.	54	
Fox, T.	77	75
Frimpong, M.		69
Gallat, S.		70
Garcia Martinez, M.	42, 133	148, 149
Geisse, G.		58
Gellynck, X.	146b	
Gereffi, G.	89	
Gibbon, P.	145	
Giovanni, D.	132	
Giraud-Héraud, E.		47, 90, 121
Gitli, E.	73o	
Goldblatt, M.	69	
Gritli, F.	73f	
Grolleau, G.	91, 92	
Grote, U.		49
Haan, C. de		4
Haddad, L.		34g
Hadi, P.U.	73k	
Harris-Pascal, C.		79
Hatanaka, M.	85	
Hattam, C.	53	
Hauselmann, P.		126
Henson, S.	4, 8 (=56), 31, 31e, 34b, 25 (=38), 39, 131, 134	1, 2, 149
Heri, S.T.	73i	
Hernández, R.		17
Holden, P.		69
Hooker, N.H.		54, 149

Author	First author reference	Other references
Hornibrook, S.		76
House, L.O.		100
Hughes, D.R.		72
Humphrey, J.	84	78, 79, 80, 89
IAAS	146	
IATRC	130	
ISEAL Alliance	116	
ITC	143	
Jaeger, S.R.		100
Jaffee, S.	1, 2, 9 (=57), 10, 16	4, 8 (=56)
Jahn, G.	114, 115	
Jano, P.		17
Jansson, A.		86
Jones, C.		48
Josling, T.	35	27
Julian, J.W.	61	
Kabwit Nguz, A.	146d	
Kanbur, R.		46
Kaplinsky, R.	29	
Keetch, D.P.	73h	
Kenny, L.		6
Kidd, A.D.	59	
Kilian, B.	48	
Kolstad I.		141
Koné Traoré, H.		34e
Kop, P. van der		19
Krissoff, B.	94	119
Kucera, D.	44, 45	
Laclau, M.		93
Ladron de Guevara, J.		58
Lamb, J.	13	
Lang, T.		117
Larsen, M.F.		30
Lheraut, G.	73c	
Loader, R.		31e, 39
Loewenther, A.		69
Lohr, L.	105, 106, 107, 119	
Loma-Ossorio Friend, E. de	73m	
López Figueroa, B.	60	
Lundy, M.		17
Lusk, J.L.	100	
Lux, L.		39
Mainville, D, Y.	122	34f
Malins, A.	71	
Manarungsan, S.	11	
Martin, A.		67, 68
Masakure, O.		16
Maskus, K.E.	23, 23a, 26	
Mathijs, E.	146c	
Mathur, V.C.	73l	
Mbaye, A.A.	12	
Meer, K van der		4
Melo, J. de		62
Michaud, J.	111	
Milestad, R.		86
Mitchell, L.	33a	
Mojduszka, E.M.	125	
Moore, M.		100
Morris, M.		29

Author	First author reference	Other references
Morrison, J.	51	126
Morrow, B.W.		100
Motte, L. de		148
Moustier P.		72
MSU Institute of International Agriculture	142	
Naewbarij, J.		11
Nelson, V.	67, 68	
Norton, G.W.	34e	
O'Brien, T.M.	55	
OECD	88	
Okado, M.	73e	
Orden, D.		27, 35
Otsuki, T.	37, 40	23a, 31d, 34c, 36
Pérez, E.		17
Perroni, C.	31c	
Pi Environmental Consulting	110	
Pick, D.	137	
Pomareda, C.	73n	
Ponte, S.	82, 83	30, 132
Poole, N.		133
Pratt, L.		48
Proforest	109	
Quinoñes, B.R.	110b	
Raikes, P.	30	
Raste, A.	110c	
Reardon, T.A.	17	22, 34f, 122, 131
Regoverning markets	21	
Rerngjakrabhet, T.		11
Roberts, D.	27, 128	35
Roekel, J. van		14
Rondot, P.	20	
Roth, E.		14
Rotherham, T.	129, 144	
Rozan, A.	99	
Sanchez, G.E.		34e, 61
Sarna, R.		44, 45
Saunders, L.		87
Schäfer, M.	73g	
Schnug, E.	146a	
Schramm, M.		114, 115
Schütz, P.		17
Serra, P.		31a
Sewadeh, M.	3	37, 40
Simo, C.	73a	
Sligh, M.	103	
Smith, G.C.	87, 113	
Soler, L-G.		47, 90, 121
Sow, M.		73a
Spackman, M.		95
Speir, J.		51
Spiller, A.		114, 115
Stenger, A.		99
Sturgeon, T.		89
Sullivan, G.H.		61
Sundkvist, A.	86	
Suwa-Eisenmann, A.		62
Swinbank, A.		39

Author	First author reference	Other references
Tallontire, A.		28, 120
Teisl, M.F.	32	
Temple, M.		148
Traill, B.		1000
Tulip, A.		59
Tumurchudur, B.		62
UK Competition Commission	81	
Umberger, W.J.	101	
UNCTAD	52, 73, 104, 127, 135, 139	
Unnevehr, L.J.	34, 34g	
Vallejo, N.	126	
Valli, C.		100
Velez, J.		13
Verbeke, W.		146b
Viaene, J.		146b
Villalobos, A.		48
Visser, M.		69
Vogel, D.	31g	
Vorley, B.	18	75, 77
Walaga, C.		59
Weldegebriel, H.T.	74	
Wigle, R.		31c
Wiig, A.	141	
Willems, S.	14	
Willer, H.	97	
Willinger, M.		99
Wilson, J.S	5 (=136), 31d, 34c, 36,	23, 23a, 26, 31, 37, 40
World Bank	Chapter 2, 43	
WTO	41	
Wynen, E.		111
Yade, M.		73a
Yussefi, M.		97
Zylbersztajn, D.		122

Part 3

Overview of operational initiatives related to private standards and trade

Introduction and notes for the reader

This chapter intends to give a concise overview of current and recent operational initiatives related to private standards and trade, with a focus on fruit and vegetable exports from developing countries. As there are numerous projects by bilateral donors and NGOs and it is not possible to know them all, this paper focuses on the activities of FAO and other international organizations. This overview is therefore not comprehensive.

1.   FOOD AND AGRICULTURE ORGANIZATION OF THE UNITED NATIONS (FAO)

FAO has standard setting activities, notably through the Codex Alimentarius Committee, but also in the fisheries sector and through various other conventions. Under the WTO agreements, national governments have an obligation to take international conventions into account when developing national regulations. Setting standards at the international level is therefore a tool for preventing the creation of barriers to trade.

However, the private sector also demands compliance with standards. Such private sector standards may be developed by buyers, by industry associations or by NGOs. Whatever the source, suppliers of food products are faced with a growing number of standards with which they have to comply.

Recognizing this reality, several units in FAO have developed activities to address issues related to private standards and market opportunities. Summary information on each initiative is given below with contact details for more information.

Food Safety and Quality Service (ESNS)

Technical assistance provided by ESNS to member countries.

Objectives: increase food quality and safety

Standards: Codex

Commodity: all

Geographical focus: global

Funding source: Regular Programme, Prevention of Food Losses Trust Fund, Trust fund for enhanced cooperation in Codex, bilaterally funded projects.

Short description and output to date:

- Policy advice on the assessment and reorganization of national food control systems for more effectiveness;
- Capacity building of food safety services: legislation, inspection and certification, laboratory facilities, quality assurance; application of GAP, GHP, GMP, HACCP, etc. Over 100 field projects dealing with different aspects of food safety.
- Training materials; practical manuals; methodologies; etc.

The scope of ESNS activities does not specifically address private standards. However, the general improvement in food quality and safety contributes to meeting requirements of private standards, the more so because in the field of food safety private standards refer often to Codex standards and guidelines.

Some specific activities that are (in)directly related to private standards and market opportunities:

- International Portal for Food Safety, Animal and Plant Health: authorized official international and national information. Link: http://www.ipfsaph.org/En/default.jsp
- Food additives database.
- Link: http://apps3.fao.org/jecfa/additive_specs/foodad-q.jsp
- Publications on HACCP, mycotoxins prevention, traceability and other related topics. Link: http://www.fao.org/es/ESN/eims_search/publications.asp?lang=en
- Project Strengthening compliance of the SPS requirements for expanded exports of fresh and processed fruits and vegetables. Geographical focus: Thailand. Contact: Mary.Kenny@fao.org

- Project Enhancement of Coffee Quality through Prevention of Mould Formation. Funded by Common Fund for Commodities.
- Contact: Renata.Clarke@fao.org
- Sub-regional Programme to facilitate trade: Food standards and food safety management. SADC member countries.
- Global Inventory Reference Materials and Food Safety Training Programme to Improve the Safety and Quality of Fresh Fruit and Vegetables. Geographical focus: Initially Latin America, expanded to Asia and Africa. Fresh Fruit & Vegetables Quality & Safety database with public and private initiatives, standards, training materials, etc.
- Link: http://www.fao.org/es/esn/fv/ffvqs?m=catalogue&i=FFVQS&p=nav; Manual. Train-the-trainers workshops, national action plans, case studies on incentives and disincentives for producers to implement quality and safety assurance programmes: 1. Cape gooseberry in Colombia; 2. Small citrus producers in Uruguay; 3. Broccoli in Ecuador.
- Contact and links: Maya.Pineiro@fao.org,
- http://www.fao.org/es/ESN/food/food_fruits_en.stm.

Animal Production and Health Division/Basic Foodstuffs Service (AGA/ESCB)
Title: Joint initiative on livestock markets, standards and market exclusion
Objective: Minimise the economic and social cost of market exclusion caused by livestock sanitary and technical standards and standard setting processes
Standards: Food safety and animal health standards
Commodity: Livestock products
Geographical focus: global
Funding source: pro-poor livestock…, Regular Programme
Short description and outputs to date:

- AGA Expert Consultation: The dynamics of sanitary and technical requirements; assisting the poor to cope. Rome, June 2004
- ESCB Symposium: Meeting international standards affecting the livestock sector: The challenge for developing countries. Winnipeg, 17 June 2004. For the Intergovernmental Group on Meat and Dairy.
- Report: The value chain approach as a tool for assessing distributional impact of standards on livestock markets: guidelines for planning a programme and designing case studies. By J. Humprey and L. Napier, IDS. January 2005.

Contact:
AGA: Anni.McLeod@fao.org
http://intranet.fao.org/en/departments/es/en/64748/73729/77807/index.html

Agricultural Management, Marketing and Finance Service-1 (AGSF-1)

Title: Cross country study on capacity building and investment needed to comply with EurepGAP standards in the Fresh Fruit and Vegetable sector
Standard: EurepGAP
   Commodity: Fresh Fruit and Vegetables
   Geographic focus: South Africa, Kenya, Chile, Malaysia
   Funding source: regular programme
   Short description:
   The study will describe the status of EurepGAP implementation in the country; organizational structures required to assist with farmer compliance; analysis of capacity building and investments needed by government, private sector and farmers to comply with and implement EurepGAP standards.
   Contact: Pilar.Santacoloma@fao.org

Agricultural Management, Marketing and Finance Service-2 (AGSF-2)

Title: Appraisal of certification costs for farmers and farmers' organizations under alternative certification schemes
   Objective: Appraise cost/benefit and managerial skills involved in organic production, marketing and certification by farmers and farmer's organizations
   Standard: organic
   Commodity: rice, fresh and processed fruits and vegetables
   Geographic focus: India, Thailand, Czech Republic, Hungary and Brazil
   Funding source: Regular Programme
   Short description:
   Case studies on successful stories of market linkages. Methodology: review of secondary information on context of production and marketing and interviews and focus group-discussion with relevant stakeholders.
   Contact: Pilar.Santacoloma@fao.org

Environment and Natural Resources Service/Organic Priority Area for Inter-disciplinary Action (SDRN/Organic PAIA)

a. UNCTAD/FAO/IFOAM International Task Force on International Harmonization and Equivalence in Organic Agriculture
   Objective: seek solutions to international trade challenges that have arisen as a result of the numerous public and private standards and regulations for organic products that now prevail worldwide.
   Standard: organic
   Commodity: all
   Geographic focus: global
   Short description
   Review the existing organic agriculture standards, regulations and conformity assessment systems including inter alia their impact on international trade in organic agriculture products. Formulate proposals on
   • Opportunities for harmonization;
   • Mechanisms for the establishment of equivalence;
   • Mechanisms for achieving mutual recognition among and between public and private systems;
   • Measures to facilitate access to organic markets, in particular by developing countries and smallholders.
Projects
   • Tunisia: Appui au Développement et à l'Organisation de l'Agriculture Biologique.
   • Objective: Améliorer les revenus et la sécurité alimentaire dans les zones

rurales à travers la diversification de la production et la valeur ajoutée que porte l'agriculture biologique.

- Croatia: Diversified Value-added Production and Certification in Environment Friendly Farming Systems.
- Objective: To improve rural income and food security through diversified production and specialty marketing of high-value and high quality products with environment enhancing production methods.
- Turkey: Formulation of a project for the development of organic agriculture and alignment of related Turkish legislation.
- Brazil: Consolidation and expansion of organic production in the Northeast of Brazil.
- Objective 2 and 3: Improved access to local, regional and international markets and access to conformity assessment systems according to market requirements.

Contact: Nadia.Scialabba@fao.org

http://www.fao.org/organicag

Raw Materials, Tropical and Horticultural Products Service-1 (ESCR-1)

Title: Socially and environmentally responsible horticulture production and trade

Objective: Provide information on niche markets that are of potential interest to developing countries and promote collaboration between certification programmes.

Standards: organic, fair-trade, Rainforest Alliance, SA8000

Commodity: fruit and vegetables and some other tropical products

Geographical focus: global

Funds: Regular Programme, CTA, ITC, World Bank

Short description and outputs to date

- Recent Market studies:
- The market for non-traditional agricultural exports. ESC technical paper 3. 2004
- The Japanese Market for Environmentally and Socially Certified Agricultural Products from Central America. 2004. IFOAM-Japan, RUTA, FAO, WB
- Environmental and social standards, certification and labelling for cash crops, Commodities and trade technical paper 2. 2003 (Includes literature review of cost/benefit analysis of standard implementation).
- Brochure ¿Es la certificación algo para mí? A practical guide to help producers and exporters in Central America with decision making about certification. Will be adapted for West Africa, East Africa and Asia.
- Regional conferences:
- Conference on Supporting the Diversification of Exports in the Caribbean/ Latin American Region through the Development of Organic Horticulture Trinidad and Tobago, October 2001.
- Production and export of organic fruit and vegetables in Asia. Proceedings of the seminar held in Bangkok, Thailand, November 2003. ESC technical paper 6, 2004.
- Working Group on Environmentally and Socially Responsible Horticulture Production and Trade and Expert Meetings. The latest working group meeting and the Fourth Expert Meeting on Voluntary Standards and Certification for Responsible Agricultural Production and Trade were held in 2004.
- Economic and Financial Comparison of organic and Conventional Citrus-growing Systems. 2001
- Increasing incomes and food security of small farmers in West and Central Africa through exports of organic and fair-trade tropical products.
- Objective: Generation of incomes and employment from increased production

and exports of organic and fair-trade products supported by effective grassroots
institutional networks. Standards: organic and fair-trade
Commodity: cocoa, pineapple, shea butter, mango Funds: German Government.
Short description: after 6 month formulation phase the project is about to start.[65]
Contact: Pascal.Liu@fao.org
Portal page: http://www.fao.org/es/esc/en/20953/20987/highlight_44152en.html

Raw Materials, Tropical and Horticultural Products Service-2 (ESCR-2)
Title: Private Standards in the US and EU fruit and vegetable markets; implications for
developing countries
Objective: Help stakeholders to understand the degree of penetration of private
standards and certification and provide policy makers with relevant information for
related national policy decisions
Standards: Private sector standards
Commodity: fruits and vegetables
Funds: FNOP
Short description:
Phase 1: Overview of existing private sector standards and relations to national
regulations and international agreements. Overview of ongoing initiatives by
international agencies that address market opportunities and constraints that result
from private standards.
Contact: Pascal.Liu@fao.org

Agricultural Sector in Economic Development Service (ESAE)
Title: Opportunities and constraints facing small farmers as a supply source for
supermarkets
Objective: Identify, and where possible, quantify barriers for linking small-scale
producers with the supermarket supply chain and other expanding markets. Facilitate
dialogue to improve the governance of chains.
Standards: not about standards per se, but those identified as barriers.
Commodity: vegetables
Geographic focus: Honduras, El Salvador
Funds: FNOP
Short description:
Strategic assessment of: i. logistics, fulfilment, quality and safety requirements of
supermarkets and other outlets; ii. Constraints faced and investments and technical
improvements needed by small farmers; iii. alternative institutional and organizational
arrangements and possible financing mechanisms to meet these requirements.
Participatory value chain analysis workshops to facilitate dialogue between chain
actors to identify bottlenecks in the chain and design action plans.
Contact: Madelon.Meijer@fao.org

---

[65] Consultant, World Bank

Agriculture Department/Priority Area for Inter-disciplinary Action (AGD/Prods PAIA)

Title: Good Agricultural Practices (GAP) activities

Objective: help developing countries cope with changing and globalizing food systems and the proliferation of GAP standards.

Commodity: all

Geographic focus: global

Funds: FNOP, Prods PAIA and regular programme

Short description:

1. GAP meta-database bubble site with a total of 853 records. Link: http://www.fao.org/wssd/SARD/sard_gap/sard-gap.htm
2. GAP workshops and e-conference:
   » International expert consultation, Rome, 2003
   » Regional FAO/IAEA GAP workshops for Asia (2003) and Africa (2004) on the application of GAP principles in the production of Fresh Fruits and Vegetables
   » Electronic conference by FAO-RLAC in 2004
   » Conference and meetings in 2004 on GAP for the livestock sector in South Africa (feed and dairy), Namibia (meat) and Tunisia (poultry).
1. Literature review of success cases where implementation of GAP and related standards helped producers improving their position in the market.
2. Projects for locally adapted GAP development in Brazil, Burkina Faso (cotton-cereal-livestock system) and Thailand (fruits and vegetables).

Contact: AnneSophie.Poisot@fao.org, http://www.fao.org/prods/GAP/gapindex_en.htm

2.   COLLABORATION AMONG INTERNATIONAL AGENCIES

Standards and Trade Development Facility

The Standards and Trade Development Facility (STDF) was established in 2002 by the World Trade Organization (WTO), the World Bank, the World Health Organization (WHO), the World Organisation for Animal Health (OIE) and FAO.

Start up funding came from the World Bank and the WTO; the latter also serves as the executing institution. Donations have been received from France, The Netherlands, Denmark, United Kingdom and Canada.

The objective of the facility is to assist developing countries in enhancing their expertise and capacity to analyse and to implement international sanitary and phytosanitary (SPS) standards, improving their human, animal and plant health situation, and thus ability to gain and maintain market access.

The facility provides grant financing for private and public organizations in developing countries seeking to comply with international SPS standards. Partner organizations may also apply for financing for projects. Project preparation grants have been approved for: CARICOM countries, Benin, Cambodia, Cameroon, Djibouti, Guinea, Mozambique, Malawi, SAARC Secretariat and Yemen. Ongoing projects include, among others, a Model Programme for Developing Food Standards within a Risk Analysis Framework and Expansion of the International Portal on Food Safety, Animal and Plant Health, both executed by FAO.

The STDF acts also as a forum for information sharing on the regular activities of the five partners: sharing of calendars, training materials and data on technical assistance.

Contact: http://www.standardsfacility.org/

In FAO: Ezzedine Boutrif, ESNS, Ezzedine.Boutrif@fao.org

The Integrated Framework

The Integrated Framework Trust Fund was created in 2001 by the International Monetary Fund (IMF), the International Trade Centre (ITC), the United Nations Conference on Trade and Development (UNCTAD), the United Nations Development Programme (UNDP), the World Bank and the WTO.

The objective if the Integrated Framework is to provide trade capacity building to governments of Least Developed Countries (LDC) and integrating trade issues into overall national development strategies.

Window I of the Fund finances Diagnostic Trade Integration Studies (DTIS). Such a study identifies sectors of export potential and supply side constraints. A DTIS also identifies measures to be taken to apply international and regional trade agreements and analyses implications for growth and poverty reduction. Recommendations are grouped into an action matrix.

Window II of the Fund provides bridging funds for capacity-building activities that are part of the DTIS Action Matrix. Funding of the Action Plan comes primarily from bilateral donors.

The WTO SPS Committee made a review of standards related issues identified in the DTIS process: G/SPS/GEN/545 28 February 2005 http://docsonline.wto.org/

Of the 14 countries that completed the DTIS, the following standards issues in the fruits and vegetables sectors were identified and follow-up action recommended:

Burundi: project for the establishment of an ISO9000 quality management scheme approved for Window II funding.

Cambodia: suggest to examine potential benefits of an export quality identity scheme tied to quality management programmes, with an accredited certifying body.

Ethiopia: training programmes in the horticultural sector should be strengthened, with attention to SPS constraints. A Window II project is being established to enhance the capacity of the Ethiopian Quality and Standards Authority.

Guinea: the United Nations Industrial Development Organization (UNIDO) has been providing assistance to the national standards and metrology institute. The national service for quality control and standards has received training and equipment to verify conformity with Codex standards.

Lesotho: recommended to adopt relevant South African legislation

Malawi: recommended that producers of groundnuts and paprika establish codes of practice, quality and other standards to prevent aflatoxin contamination in order to be able to diversify exports.

Mozambique: need to update pest status report and need help in eradicating or isolating some pests and in meeting other SPS requirements.

Nepal: need enforcement of quality standards and SPS requirements at farm and processing stages if Nepal wants to increase agricultural exports.

Senegal: a monitoring system in the groundnut sector is required to ensure quality and prevent aflatoxin contamination.

Yemen: Yemen Standards, Metrology and Quality Control Organization has been established with assistance from UNDP and UNIDO in the late 1990s. Further assistance on standards through a Window II project is being provided by UNIDO and a Window II fruit and vegetable export promotion project is implemented by UNIDO and FAO.

Unfortunately the IF web site does not give any information on the above mentioned approved or ongoing Window II projects.

Contact: http://www.integratedframework.org/

3.   WORLD BANK
Standards and Trade E-learning Course

This course was provided in April-May 2005 to 200 participants from 65 countries. It was targeted to policy makers, staff of standards bodies and research institutes and managers of companies in developing countries, staff of supporting agencies and NGOs.

The objective was to make participants more aware about the opportunities and risks, costs and benefits and development potential associated with the application of internationally recognized product and process standards

The course examined the political economy dynamics of standards-setting, application, and enforcement at international and national levels, with a particular emphasis on the position of developing countries. The course provided documentation plus online discussion. At the end, participants were required to prepare a Response Action Plan to a problem related to standards and trade in their country of employment.

Contact: Steven Jaffee, sjaffee@worldbank.org

http://web.worldbank.org/WBSITE/EXTERNAL/WBI/WBIPROGRAMS/TRADELP/0,,contentMDK:20289384~menuPK:461730~pagePK:64156158~piPK:64152884~theSitePK:461702,00.html

4.   INTERNATIONAL TRADE CENTRE (ITC)

The ITC was established in 1964 and is jointly operated by the WTO and UNCTAD. The Centre provides information on export markets and marketing techniques and assists in establishing export promotion programmes and marketing services. The Centre's help is freely available to the least-developed countries.

Joint integrated technical assistance programme

This ITC/UNCTAD/WTO Common Trust Fund receives funds from 13 donors and provides capacity building to government officials of developing countries on the WTO agreements since 1998. Recently this included two regional workshops on the SPS agreement in Zambia and Mali and two regional workshops on the TBT agreement in Malawi and Cotonou.

Contact: http://www.jitap.org/

Standards and Quality Management programme

This programme assists enterprises and organizations dealing with Standardization, Quality Assurance, Accreditation and Metrology (SQAM) in developing countries. Its work comprises all standards, also in non-food sectors, but the focus is on quality standards.

Training materials and information products that may be of interest to the fruit and vegetable sector include:

» Training packs on SPS and TBT: The WTO/TBT Agreement; a business perspective.
» ISO 9001 Fitness Checker: CD-ROM sold through national trade support organizations, with a question & answer mechanism that guides enterprises through requirements and checks what still needs to be done for certification.
» World Directory of Information Sources on Standards, Conformity Assessment, Accreditation, Metrology, Technical Regulations, and Sanitary and Phytosanitary Measures: Includes standards bodies, TBT/SPS enquiry points, legal metrology organizations, laboratories and accreditation bodies; contact points for Codex Alimentarius Commission, OIE and IPPC.

Handbooks:

Export Quality Management: An Answer Book for Small and Medium Sized Exporters. 2001 (+ national adaptations with partner organizations in Argentina, Bangladesh, Brazil, China, Korea, Malaysia, Nigeria and Central Asia)

ISO/ITC 2002. ISO 9001 for Small Businesses: What to do. Revised edition

Contact: http://www.intracen.org/eqm/

International Trade Centre/Commonwealth Secretariat (ITC/ComSec): Influencing and Meeting International Standards

This is a joint initiative of ITC with the Commonwealth Secretariat, implemented in the period 2003-2005. Its objective is the identification of training needs and development of project proposals for the implementation of the WTO TBT and SPS agreements and participation in the relevant WTO committees.

Case studies were conducted in 2003 in Jamaica, Kenya, Malaysia, Mauritius, Namibia and Uganda that have resulted in a publication with recommendations for technical assistance (see overview of analytical work). Workshops were held in Cairo and Bangladesh on the SPS agreement and in Kyrgyzstan, Tajikistan and Kenya on the TBT-agreement. An international workshop with donor countries was held in Geneva 2005 to review Mentoring and Twinning Arrangement recommendations and to develop project proposals.

Contact: http://www.intracen.org/eqm/pages/geneva_workshop_june05.htm

ITC Bolivia: Promotion and expansion of exports of selected products

This project started in 2001 and is implemented by ITC in collaboration with SIPPO and the Instituto Boliviano de Comercio Exterior (IBCE), the Centro de Promoción-Bolivia, C-PROBOL and Cámara de Exportadores de Santa Cruz (CADEX). The project is funded by the Swiss Government.

The project's objectives are to increase export capacity and exports, including exports of nuts, beans and organic products. The module on quality management provides assistance in the areas of ISO/IEC 17025, HACCP and ISO 9000.

Contact: www.exportebolivia.net

5.   UNITED NATIONS CONFERENCE ON TRADE AND DEVELOPMENT (UNCTAD)

Most of UNCTAD's work on standards is in the area of environmental standards under its Trade, Environment and Development Programme.
http://r0.unctad.org/trade_env/test1/openF1.htm

Consultative Task Force on Environmental Requirements and Market Access

The Consultative Task Force (CTF) was inaugurated in June 2004 and held its first substantial meeting in November 2004. The CTF is funded by the Netherlands. Its objective is to assist developing countries in analyzing key trends of environmental requirements in export markets. Furthermore, the CTF seeks to exchange national experience on pro-active approaches to meeting these environmental requirements with a view to maintaining market access, harnessing developmental gains and safeguarding social welfare.

The CTF works through open-ended multi-stakeholder forums. The first meeting recommended that a working group of CBI (NL), FAO, Inmetro (Brazil), and others would be set up to comment on a draft feasibility study prepared by UNCTAD on the possible added value of an internet portal on environmental and related health requirements for market access. The meeting also recommended that projects should be developed to assist developing countries to develop national GAP to be benchmarked against EurepGAP.

Contact: http://www.unctad.org/en/docs//ditcted20052_en.pdf

Capacity-Building Task Force for Trade, Environment and Development

This Task Force was set up in 2000 by UNCTAD and the United Nations Environmental Programme (UNEP). Its objective is to strengthen the capacities of developing countries and countries with economies in transition, to effectively address trade-environment-development issues. It has developed many activities, many not related to standards.

However, a Policy Dialogue on Promoting Production and Trading Opportunities for Organic Agricultural Products was organized in 2002 in Brussels. This dialogue has resulted in a programme on Promoting Production and Trading Opportunities for Organic Agricultural Products in East Africa that is planned to start in August 2005.

Contact: http://www.unep-unctad.org/cbtf/

At UNEP: Asad Naqvi at UNEP, asad.naqvi@unep.ch, at UNCTAD: Nuria Castells, nuria.castells@unctad.org

UNCTAD-FIELD II

UNCTAD-FIELD II or the: Building Capacity for Improved Policy Making and Negotiation on Key Trade and Environment Issues is a follow-up to the UNCTAD-FIELD I project. It is implemented by UNCTAD and the Foundation for International Environmental Law and Development (FIELD). The planning phase was executed in 2002 and implementation of the project started in 2003. The project has a budget of one million pounds sterling and is funded by DFID.

The project is implemented in 15 developing countries in three regions, with one core country in each region (South-east Asia, East and Southern Africa and Central America and the Caribbean). For continuity there is a partly overlap with target countries of UNCTAD-FIELD I project. Each region was asked to select two key issues.

The Latin American countries (Costa Rica, Cuba, Guatemala, Honduras, Nicaragua, Panama and the Dominican Republic) concentrate on environmental requirements and market access, and especially on organic agriculture and organic markets.

South-east Asian countries selected the theme of linkages between environmental requirements and market access, including their export competitiveness. Horticulture is among the focus product groups for Bangladesh, Cambodia, China, Philippines, and Viet Nam. No activities are reported for Africa.

Contact: Project manager Latin America: Nuria Castells, nuria.castells@unctad.org

Analysis of trade implications of ISO 14001 and training

In 1997 UNCTAD organized an Expert Meeting to examine the trade implications for developing countries of international standards for environmental management systems. Follow-up analysis was funded by Italy and the Netherlands. The results have been incorporated into the current training package on trade, environment and development (also available online) in its module 6 on Implications of international standards for environmental management systems, particularly ISO 14000, for developing countries and countries in transition.

Furthermore, a trade policy capacity building project for India has also dealt with trade implications of ISO 14001 and possible strategies to promote the use of environmental management systems by the Indian industry.

Contact: meeting and analysis: rene.vossenaar@unctad.org, training: Nuria Castells, nuria.castells@unctad.org

Reconciliation of Environmental and Trade Policies

This project, funded by the Netherlands, was implemented by UNCTAD in cooperation with UNDP and the German Association for Technical Co-operation (GTZ) in 1999 and 2000.

It was divided in two clusters and the second cluster concentrated on promoting the export of environmentally preferable products of developing countries.

The objective was to:

- encourage company or (industry) association partnerships between eco-pioneers in developed and developing countries along the supply chain;
- facilitate access to information on and review of existing and emerging environmental requirements in target markets;
- facilitate the training of managers on cleaner production methods and products; and
- encourage exporting firms in using environmental management systems, such as ISO 14001.

A bilateral workshop was organized between Indian exporters, export promotion boards and Government bodies and German importers, wholesalers, industry associations and Government institutions for enhancing export of organic food and beverages from India to Germany. Another seminar was organized in 2000 at BioFach. Contact: Ulrich Hoffmann. ulrich.hoffmann@unctad.org

6.   WORLD TRADE ORGANIZATION (WTO)

The WTO regularly organizes courses for government officials from developing countries and economies in transition.

- WTO Trade Policy Courses: 3 months duration
- Specialized course on SPS measures in Spanish and e-learning modules on the SPS Agreement
- Regional workshop on the SPS Agreement in Jamaica, in collaboration with the IADB, CARICOM and the Government of Jamaica

http://www.wto.org/english/tratop_e/devel_e/train_e/tradepolicycourse_e.htm

The WTO has also created over 100 computerized reference centres in trade ministries in developing countries and in headquarters of regional coordinating organizations.

7.   UNITED NATIONS DEVELOPMENT PROGRAMME (UNDP)

Time did not permit to make a comprehensive overview of all UNDP projects that have a component on standards. An example of such a project is Rural Enterprise Development Programme in Bhutan, which UNDP supports in collaboration with SNV, FAO and UNIDO. This programme is implemented from 2002 to 2007 to build a positive environment for rural enterprise and take advantage of business opportunities for niche products. This includes developing a management information system, and formulating standards and quality control guidelines. In addition, a formulation mission funded by the EU in the spring of 2005 was to formulate an agricultural production project with a certification component.

Contact: http://www.undp.org.bt/poverty.php

8.   REGIONAL ORGANIZATIONS

Inter-American Institute for Cooperation on Agriculture (IICA) Agriculture Health and Food Safety Directorate

Agriculture health & Food safety is an important theme for the Inter-American Institute for Cooperation on Agriculture (IICA). IICA supports the development of regulatory mechanisms, science-based technical capacities, and institutional infrastructure. The organization also supports the implementation of the SPS Agreement and of decisions of the OIE, IPPC and Codex. Furthermore, IICA assists in the improvement of national agricultural health and food safety services and in the development of food safety capacity throughout the food chain.

Since 2003, IICA and USDA implement the Initiative on Strengthening institutional capabilities in the field of SPS measures, funded by USDA. This initiative aims to promote the presence and active participation of capital-based experts at meetings of the SPS Committee.

Contact: IICA provides the Infoagro.net web site with very up-to-date information on food safety under its agrosalud heading: http://infoagro.net

United Nations Economic and Social Commission for Asia and the Pacific (UNESCAP)

The UN Economic and Social Commission for Asia and the Pacific (UNESCAP) has developed a programme on Marketing Green and Organic Agricultural Produce as a Tool for Rural Poverty Alleviation. The programme is implemented since 2000 and targets governmental, NGO and research institutes in Asia and the Pacific.

The programme's objectives are:

- to highlight the positive link between effective marketing of green and organic produce and rural poverty alleviation;
- to enhance the capacity of governmental and non-governmental organizations in Asia to design and build marketing channels for organic and green produce

by disseminating existing successful practices;
- to facilitate structured information flows.

Under the programme UNESCAP has initiated the Asia-Pacific Information Network on Organic Farming and Green Food (OFGF) to strengthen capacities to promote and practice organic farming and sustainable agriculture. The network is facilitated by the China Green Food Association and has seven member countries.

The network provides technical advisory services, training, study tours, information and staff exchange. Country reports have been prepared. A symposium on pro-poor certification systems for green and organic produce was organized in China in June 2005.

Contact at UNESCAP: Yap Kioe Sheng yap.unescap@un.org, Margot Schuerman,

schuerman.unescap@un.org   http://www.unescap.org/pdd/prs/ProjectActivities/ Ongoing/GreenOrganicTool/green_organic_tool.asp

OFGF: http://www.ofgf.net/index.htm (Out of order since first visit, please see reference http://www.unescap.org/pdd/prs/organic.asp)

Asia-Pacific Economic Cooperation Council (APEC)

The Agricultural Technical Cooperation Working Group (ATCWG) of APEC is organizing various standards-related workshops in Bangkok. A workshop on supply chain management in August 2005 is to improve university-industry linkages and to propose action plans with a focus on quality standards in international trade. This is to be followed by a workshop on quality of fresh produce in September 2005.

Furthermore, in 2005 the ATCWG approved a work plan that includes the development of a network system on grade standards, requirements and regulations and capacity building activities on quarantine measures, the SPS agreement, pest risk analysis and the ISPM 15 standard on wood packaging.

Contact: http://www.apec.org/apec/apec_groups/working_groups/agricultural_ technical.html

Southern African Development Community (SADC)

The Food Agriculture and Natural Resources Department (FANR) of the Southern African Development Community (SADC) is reviewing the new EU pesticide MRLs and preparing a project with COLE-ACP for capacity building.

Contact: http://www.sadc.int/index.php?action=a1001&page_id=fanr_agriculture

The Southern African Development Community Accreditation (SADCA) is the regional accreditation structure. The SADCA mandate is to define a suitable accreditation structure for the region and facilitate the creation of a pool of internationally acceptable accredited laboratories and certification bodies.

Contact: http://www.sadca.org/, http://www.sadc-sqam.org/

Association of Southeast Asian Nations (ASEAN)

Within the framework of the Association of South East Asian Nations (ASEAN), a proposal for a regional GAP was developed in 2001. In 2003 the project, funded by ASEAN and AUSAid, was redesigned with inputs from the Ministries of Agriculture. The project has developed a compendium on ASEAN Quality Standards and a compendium on SPS requirements. An ASEAN-GAP is under development. A workshop to finalise the GAP will be held in Singapore in November 2005. For 2006 train-the trainer workshops are planned. Pre-training activities have taken place in Cambodia, Myanmar, Lao People's Democratic Republic and Viet Nam, to enable them to fully participate in the GAP development.

Contact: www.aphnet.org

9.   BILATERAL ASSISTANCE

European Information System for Organic Markets (EISfOM)

This EU funded programme intends to establish extensive databases with information on organic markets. However, so far only data on organic production in the European Union are available. As a first step, an inventory of existing organic market data collection systems at the national level has been made. In November 2005 EISfOM held a seminar entitled Towards an European Framework for Organic Market Information. It may therefore be expected that in the future EISfOM will publish market data that are of relevance to producers and exporters in developing countries.

Contact: http://www.eisfom.org/

Centre for the Promotion of Imports from Developing Countries (CBI)

The Centre for the Promotion of Imports from Developing Countries (CBI) was established in 1971 and is an agency of the Dutch Ministry of Foreign Affairs. The centre seeks to contribute to the economic development of developing countries by strengthening the competitiveness of companies from those countries on the EU market. CBI considers social values and compliance with the most relevant environmental requirements to be an integral part of its policy and activities.

The CBI offers the following programmes:
- Market information:
  CBI's database on market information and non-tariff trade barriers on more than 50 product groups, including market surveys, strategic marketing guides and market entry and export planning manuals. The CBI Access Guide provides a very user-friendly and up-to-date database on all the EU regulations and private sector and voluntary standards. Registration for companies and business support organizations in developing countries is free of charge. (http://www.cbi.nl/accessguide/)
- Company matching:
  The company matching programme links well-versed suppliers in developing countries to reliable importing companies in the European Union and vice versa
- Export Development Programmes (EDPs):
  EDPs assist entrepreneurs in developing countries in entering and seeding on the EU market and/or consolidating or expanding their existing market share. This includes compliance with regulations and private standards and other market requirements. The EDP on Fresh Fruits and Vegetables is currently being implemented in Egypt, Jordan, Kenya, Senegal, and Uganda. Contact: Peter van Gilst (31)10-2013415. The EDP on organic food ingredients for industrial use is being implemented in Bolivia, Colombia, Cuba, Ecuador, Egypt, Ethiopia, India, Indonesia, Pakistan, Peru, Philippines, Sri Lanka, Tanzania, Uganda and Zambia. Contact: Cor Dieleman (31)10-2013423.
- Training Programmes for exporters and business support organizations.
- Business Support Organizations development programme

Contact: http://www.cbi.nl/, cbi@cbi.nl

Swiss Import Promotion Programme (SIPPO)

The Swiss Import Promotion Programme (SIPPO) was launched in 1982 and operates under the Swiss State Secretariat for Economic Affairs. SIPPO helps small and medium-sized enterprises in emerging markets and markets in transition to enter the Swiss and European Union market, and provides Swiss importers with assistance in finding new products, new suppliers and new sourcing markets.

SIPPO offers:
- Trade and market information
- SIPPO produces market overviews and information on European norms, standards, quality requirements and import regulations, customs regulations and charges. Examples of publications:
- SIPPO and FIBL. 2001 (2004). The Organic Market in Switzerland and the

European Union. Zürich/Frick, January 2001, 2nd Edition: 2004.
- SIPPO. 2001 (2002) Fruits and Vegetables / Switzerland (With a large section on regulations and buyers' requirements)
- SIPPO. 2004. Fruits and Vegetables / Macedonia
- SIPPO and CBI. 2005. Exporting to Switzerland and the European Union
- Available at: http://www.sippo.ch/cgi/news/publications.asp?mode=6#agrimc
- Trade development programme:
- Support in design, quality management and product adjustment and sales promotion measures such as mailshots and trade fair attendance. SIPPO also examines cooperation with Swiss companies in a private-public-partnership.
- As part of its Fruits and vegetable programme, companies from Bulgaria, Ghana, Ecuador and Viet Nam will participate in Fruit Logistica fair in 2006
- As part of its Organic products programme, companies from Bolivia, Bosnia and Herzegovina, Bulgaria, Egypt, Ghana and Peru will participate in BioFach 2006
- Trade promotion
- SIPPO attempts to find exportable products for which there is a demand in the Swiss and the EU markets and supports the establishment of contact with new suppliers (sourcing).
- Training programmes on: export marketing; export administration and logistics; process-oriented marketing; Export Quality Oriented Enterprise Profiling Project; ISO 9000 / 9001; HACCP; organic production; and export promotion tools for the Swiss market.

Contacts: www.sippo.ch, info@sippo.ch

United States Agency for International Development (USAID)

The USAID Trade for African Development and Enterprise (TRADE) initiative seeks to help African countries to make better use of opportunities offered by the African Growth and Opportunity Act (AGOA).

The initiative is opening several TRADE hubs that undertake various standards related activities.

West African Trade Hub

The West African Trade Hub (WATH) has its office in Ghana, but another office just opened in Senegal. WATH is working with the Economic Community of West African States (ECOWAS) and the Union Economique et Monétaire Ouest Africaine (UEMOA) to harmonize West African sanitary and phytosanitary standards. And it will provide training to officials to participate fully in World Trade Organization talks. Furthermore, it liaises with US Agencies and international organizations to train export-ready West African companies and public institutions in SPS requirements for exports into the United States.

Contacts: Kofi Humado at khumado@watradehub.com http://www.watradehub.com/

East and Central African Trade Hub

The East and Central African Trade Hub (ECATrade Hub) has its office in Kenya. The Hub plans to work with the private sector for a more effective use of international trade standards, ecolabelling and ISO environmental management system standards.

The hub has published the following publications relating to standards:
- ECA Trade Hub. 2005. All you have to know about the European Community's Rules of Origin
- William Hargraves, ACDI-VOCA. 2005. Survey of Existing and Planned SPS/Food Safety Standards Activities in the COMESA Region.

Contact: http://www.ecatradehub.com/home/index.asp

Southern African Trade Hub

Within the Southern African Trade Hub standards related activities mostly take place under the Trade Competitiveness Component. Among other activities, this component works with the horticulture and cotton supply chains and seeks to improve the ability of firms to deal with emerging quality standards. With regard to increasing market access, baby corn and baby carrots from Zambia, table grapes from Namibia, and litchis from South Africa are the first targets for new admissibility into the United States. In the paprika sector, the TRADE will focus on helping industry to gain HACCP accreditation and international certification for good laboratory practices and testing.

Contact: http://www.satradehub.org/

Other examples of bilaterally funded assistance projects

It was not possible to track all bilaterally funded agricultural projects with a standards or certification component. Below are just three examples of such projects.

TechnoServe: organic pineapple exports from smallholder producers in Ghana. TechnoServe is a non-profit organization, founded in 1968 in the United States, to provide the rural poor of the developing world with the technologies they needed to improve their productivity. By the early 1970s, however, the organization evolved to focus on community-based, small-business development. Since 1998 TechnoServe serves business of all shapes and sizes and draws staff from the private sector.

TechnoServe has been providing technical assistance to Ghanaian pineapple farmers and processors since 1993. One of TechnoServe's clients is Athena Foods Limited, a company that processes pineapples and citrus fruits into concentrates and fruit juices for local and export markets. Athena liked to tap into the growing and more lucrative organic food market. TechnoServe linked Athena to 311 small-scale organic pineapple and citrus farmers whose only buyers had been local market women. TechnoServe helped the farmers to prepare the documentation, institute the internal control systems and get the technical training they needed to receive their organic certification.

http://www.technoserve.org/africa/ghana-pineapple.html

CLUSA: Organically grown and certified coffee, cacao, sesame and cashew to markets in the United States, Europe and Japan and in sales and distribution of organic vegetables and other commodities to local restaurants, hotels and supermarkets in El Salvador.

Organized in 1916 as the Cooperative League of the USA (CLUSA), the National Cooperative Business Association (NCBA) is still recognized in many countries under the CLUSA name. NCBA's CLUSA International Program began providing assistance to developing countries in 1953.

In El Salvador, through the Non-Traditional Agricultural Exports (NTAE) Production and Marketing Project, CLUSA worked with 128 cooperatives and 6 secondary organizations. The project pioneered in organic production, in exporting organically grown and certified coffee, cacao, sesame and cashew to markets in the United States, Europe and Japan and in sales and distribution of organic vegetables and other commodities to local restaurants, hotels and supermarkets. The project benefited 60 000 producers and their families. Two farmer organizations involved in the project won the country's first major environmental awards for their organic production work. CLUSA assisted to launch PROEXSAL, an import/export cooperative owned by farmer cooperatives and grower groups with some exporters. PROEXSAL provides quality control for horticultural crops shipped from the rural areas to the city for sale and has imported and sold US and Guatemalan horticultural commodities. It continues to market fruits and vegetables grown by Guatemalan and Salvadoran farmers, 80 percent of which are organically produced, to restaurants, hotels and supermarkets.

http://www.ncba.coop/clusa.cfm

ICCO and Agrofair Assistance & Development: Development of new fruit sources that can be marketed in Europe under the Fairtrade label organic fruit.

AgroFair Ltd. is an importer and distributor of Fairtrade labelled and organic tropical fresh fruit. Agrofair is based in the Netherlands but its fruit is sold in various European countries. The company is co-owned by its producers. Agrofair Assistance & Development is an NGO and works in close cooperation with AgroFair Ltd. It develops a wide range of fair-trade and organic fruit produce by assisting conventional producers with conversion and guiding producers through different kinds of certification processes in various areas such as social (e.g. FLO, ETI and SA8000), environmental (e.g. BCS, Skal and Ecocert) and technical (e.g. EurepGAP, ISO, HACCP) certifications.

ICCO (Interchurch organization for development co-operation) is based in the Netherlands and receives funds from the Dutch Government. ICCO provides funds to Agrofair Assistance and Development and ICCO will provide further support in the field of sales and certification and possibly engage in lobbying activities with regard to trade barriers (bananas, citrus). In the past ICCO has also supported fair-trade banana producers in the Dominican Republic.

www.icco.nl/english and http://www.agrofair.nl

10.   PUBLIC-PRIVATE PARTNERSHIPS
Sustainable Commodities Initiative

This is an initiative of UNCTAD and the International Institute for Sustainable Development (IISD), an independent, non-profit company. The umbrella initiative is funded by the IDRC/CDRI of Canada and has the objective to improve the social, environmental and economic sustainability of commodities production and trade by developing global multi-stakeholder strategies on a sector-by-sector basis.

The initiative has up till now focussed on coffee and meetings and workshops have led to the establishment of the Sustainable Coffee Partnership with its own programme and funding structure.

Contact: cwunderlich@iisd.ca

The Global Food Network

The Global Food Network was initiated by the EU Commission Scientific Officer and its activities are facilitated by Agri Chain Competence Centre (ACC), a consultancy company in the Netherlands. It is essentially a network of research institutes from Europe (Portugal, United Kingdom, Spain, Netherlands, Denmark and Hungary), MERCOSUR (Argentina, Brazil, Paraguay and Uruguay) and ACP countries (Uganda, Kenya and the Caribbean).

The Network was started in 2002 and will be operational at least until 2005 in order to improve the cooperation in research on food safety and food quality between EU countries, ACP and MERCOSUR. Activities include an inventory of public and private quality and safety standards, conferences, electronic discussions, reports and internationally attuned research agenda's

Contact: http://www.globalfoodnetwork.org info@globalfoodnetwork.org.
ACCC: (31)735286659

The Food Quality Schemes Project

This project is implemented by the Institute for Prospective Technological Studies of the EC Joint Research Centre in cooperation with DG Agri and DG RTD and the University of Bologna. Funds come from the European Union. The project is planned to run from 2005 to 2007. The project will analyse the opportunity to implement a Community-wide legal framework for protection of Quality Assurance and Certification Schemes in the Food Chain. Components of the project will be a research study and stakeholder workshops. This will lead to a report followed by a stakeholders' hearing to discuss the outcomes and a final conference in 2006/07.

Contact: jrc-ipts-foodquality@cec.eu.int

Sustainable Trade and Innovation Centre (STIC)

This initiative was launched in Brussels in 2002 by the European Partners for the Environment (EPE), Commonwealth Science Council and Royal Tropical Institute (KIT). The objective was to establish a global partnership for information exchange, innovation and forging business partnerships. However, no follow-up activities have been reported since.

Contact: http://www.epe.be/euhub/

Sustainable Food Laboratory

General

The Sustainable Food Laboratory is a collaborative initiative of companies, governmental institutes and civil society organizations. Participating companies are Carrefour, General Mills, Nutreco, Organic Valley Cooperative, Rabobank, Sadia, Sodexho, Starbucks, SYSCO, and Unilever. Governmental institutes come from Brazil, the Netherlands, the European Commission and the International Finance Corporation (IFC), and the World Bank also participate. Civil society is represented by Consumers International, Oxfam, The Nature Conservancy, and the World Wide Fund for Nature(WWF).

The Secretariat of the Lab is formed by Generon consulting, Sustainability Institute (a non-profit consulting group) and the Synergos Institute (non-profit, Global Leadership Initiative). The Lab is planned to function from 2004 to 2006 with the objective to make food systems more economically, environmentally, and socially sustainable.

Funding comes from participating organizations and from Shell Foundation, W.K. Kellogg Foundation, Charles Leopold Mayer Foundation and King Baudouin Foundation. The Lab secretariat has a budget of US$3.2 million, of which US$1.5 million remains to be raised. Funds for separate pilot initiatives of the Innovation Teams have also to be raised.

Responsible Commodities Initiative

One of the lab's sub-projects is the Responsible Commodities Initiative. Participating organizations are the Brazil Specialty Coffee Association; Ethical Certification and Labeling Space, IFC, Medley Global Advisors; Innovation Network, Min. of Agriculture NL; Rabobank International; Rainforest Alliance; The Nature Conservancy; Unilever and WWF.

Its objective is to mainstream sustainable business practices in food chains, especially in palm oil, cotton, soy, sugar and coffee chains.

Planned products:
- Commodity Meta-Standard (~ matrix to "value" existing standards)
- Commodity Purchasing screens (~ standard for sustainable procurement)
- "Equator Principles" for commodity finance (~ standard for sustainable investment)
- Shareholder activist information pack
- Commodity futures screens (It is not clear to the author what this means)

The Business Coalition for More Sustainable Food

Another subproject of the Sustainable Food Laboratory is The Business Coalition for More Sustainable Food (BCSF), to harness the buying power of food-related companies to create more sustainable food supply systems. The BCSF plans to: create a clearing house of best practices for production and marketing; coordinate opportunities for common purchase specifications that improve economics and enhance sustainability; initiate collaborative projects that solve problems up and down food chains; and to provide a safe space for collaboration with leaders from academia, non-profit organizations and government.

Latin America family farms

The subproject Latin America family farms works with farmers in the Dominican Republic, Guatemala, Brazil on innovation of food supply chains, including cost-benefit analysis of standards. The subproject will also facilitate learning across the hundreds of existing projects that address market access and sustainability of family farms.

Contact: http://www.glifood.org/

Hal Hamilton, Executive Director, Sustainability Institute, hhamilton@sustainer.org, +1 (802) 436-1277 x101

Adam Kahane, Director, Generon Public Service, kahane@generonconsulting.com.org, +1(978) 232-3500 x30

Ethical Certification and Labeling (ECL) Space

The ECL Space is implemented/facilitated by Pi Environmental Consulting. The steering group is formed by RBF/Yale University, UNCTAD, ISEAL Alliance, Consumers International, WWF, Imaflora, ScanComGroup Viet Nam, Fern/UICN, Asian coalition for SME development, Icontec and Unilever.

The ECL Space aims to solve problems that limit the acceptance and effectiveness of environmental and social certification and labelling. It conducted background analysis on different typologies of certification scheme and impact on market access in 2003 and 2004. A follow-up work programme is not yet published or has still to be decided.

Contact: Pierre Hauselmann phauselm@piec.org and Nancy Vallejo nvallejo@piec.org. http://www.piec.org/ecl_space/

The Working Group on Standards and Trade

In 2005 the World Bank took the initiative to create a Standards Working Group, to be facilitated by Chemonics International, a consultancy company. An inaugural meeting was held in June 2005 with participation of several multilateral and bilateral agencies, non-governmental organizations and consultancy companies.66 These organizations are also the potential members of the Standards Working Group.

The working group is intended to be a knowledge network on agro-food standards, but the precise mission, objectives and activities have still to be decided upon.

Contact: Steven Jaffee, sjaffee@worldbank.org, Matthew Edwardsen medwardsen@chemonics.com

11.   NGO ALLIANCES

ISEAL Alliance

The International Social and Environmental Accreditation and Labelling (ISEAL) Alliance was established in 2000. Its members are FLO, FSC, IFOAM, IOAS, MAC, MSC, SAI, SAN/Rainforest Alliance, associate members: GEN, Chemonics. Funds come from GTZ, Overbrook Foundation and HIVOS Biodiversity Fund. The ISEAL Alliance mission is helping to create a world where ecological sustainability and social justice are the normal conditions of business.

The ISEAL Alliance activities are:
- Peer review for standard setting and for accreditation practices
- Social Accountability in Sustainable Agriculture project: to learn how to audit

---

66 Agencies: FAO, UNCTAD, UNIDO, the European Commission (EC), Pesticide Initiative Programme (COLEACP-PIP), United States Agency for International Development (USAID), InterAmerican Institute for Cooperation in Agriculture (IICA), GTZ

NGOs: ISEAL Alliance, Rainforest Alliance, World Wildlife Fund (WWF), Oxfam, International institute for Sustainable development (IISD)

Consultancies: Chemonics International, Abt Associates, DAI, and the Institute for Food and Agricultural Standards at Michigan State University (MSU).

and certify against social and labour standards given the diverse social, cultural and economic situations in agriculture (2000–2004)

- Developed Code of good practice for setting social and environmental standards
- Code of ethics for ISEAL members
- Advocacy
- Services to members (e.g. tenders for external auditing against ISO norms)

Contact: http://www.isealalliance.org; Patrick Mallet pmallet@isealalliance.org, Sasha Courville, secretariat@isealalliance.org

The Pacific Institute and the International NGO Network on ISO (INNI)

The Pacific Institute was founded in 1987 and is based in Oakland, California, to provide independent research and policy analysis on issues at the intersection of development, the environment and security. Its Economic Globalization and the Environment (EGE) Program studies the effects of the increasingly integrated global economy on the environment and society.

As a member of the US Technical Advisory Groups, the EGE programme attends the ISO technical committees on issues including water management, ecolabelling and environmental communications. The Pacific Institute was a founding member of the NGO Working Group on ISO 14000, which worked to strengthen the voice of non-governmental organizations (NGOs) participating in the standards-setting process, and EGE programme director Jason Morrison currently serves as Chair of the ISO/TC 207 NGO Task Group.

Building on the ISO14001 experience, the Pacific Institute set up the International NGO Network on ISO (INNI), around 2003/2004. Active membership is limited to non-governmental organizations working on issues such as environmental management, climate change, water management, and corporate social responsibility, among others. Observer membership is open.

The objective of INNI is to ensure that any ISO-created environmental standard serves the public interest and protect the environment. By providing timely information on the activities of ISO to network organizations they can activate their constituents, provide guidance to decision-makers, and shape public opinion.

Contact: http://www.pacinst.org/ and http://inni.pacinst.org/inni/index.htm
Jason Morrison: jmorrison@pacinst.org

RAFI-USA Just Food programme

The Rural Advancement Foundation International (RAFI)-USA traces its heritage to the National Sharecroppers' Fund, which was founded in the 1930s. RAFI-USA concentrates on North Carolina and the south-eastern United States, but works also nationally and internationally. Its aim is a reliable supply of safe, healthy food by strong family farms and rural communities, and with close connections between consumers and producers, environmentally sound farming and safeguarding of agricultural biodiversity.

Under its Just Food programme it develops activities in the area of organic farming and organic, environmental, and fair-trade standards and labels that give small and modest-size farmers marketplace rewards.

RAFI drafted social justice standards for organic agriculture that they presented at a social justice workshop in 2003 in Canada at the IFOAM World Conference. They organized another social justice workshop with a broader agenda in 2004 in Thailand in connection with the IFOAM Organic Trade Conference. http://www.rafiusa.org/index.html

12. NATIONAL CERTIFICATION PROGRAMMES FOR BENCHMARKING

A distinct type of initiatives to address standard-related marketing constraints and opportunities are national programmes that are developed with the specific purpose of being recognized as equivalent to certification systems in import markets.

There are mainly two categories: organic regulations and national GAP programmes

National GAP programmes

ChileGAP: http://www.buenaspracticas.cl
Peru: http://portalagrario.gob.pe/dgpa_prac_agricolas.shtml
Mexico: http://www.mexicocalidadsuprema.com
Malaysia: farm accreditation scheme of Malaysia
http://agrolink.moa.my/doa/SkimAkreditasiLadangMalaysia.htm
China: http://www.cnca.gov.cn/ (Chinese only), http://www.eurep.org
ASEAN: www.aphnet.org

National organic programmes

The developing countries with national organic programmes that have been recognized as equivalent to the EU organic guarantee system are:

Argentina: http://www.infoagro.com/agricultura_ecologica/ecologia_argentina/NORMAS/normas.asp Costa Rica: http://www.infoagro.go.cr/organico/PLAN_ACCION_2000.doc

For other national organic programmes and legislation, please see the FAO Organic Agriculture Information Management System (Organic-AIMS):

http://www.fao.org/organicag/frame6-e.htm

Conclusions and Recommendations

1. Discussion and conclusions on standards and trade

The first part of this report has presented an overview of international agreements, national regulations and private standards. Standards have been examined by category: general quality, traceability and origin labelling standards; food safety standards; sustainable agriculture and Good Agricultural Practices; environmental standards; organic agriculture; and labour and social standards. The second part of the report has reviewed the literature on standards and trade. Finally, a summary of major operational initiatives has been presented.

The following general conclusions can be drawn from the above analyses.

Phytosanitary standards are only found in the regulatory domain – No private phytosanitary standards were found.

This is not surprising, as phytosanitary issues have long been regarded to fall in the public domain. Problems caused by imported plant diseases or pests may affect a country's economy and natural resources. However, the firms and persons involved in the movement of goods and the persons that may accidentally import these pests are generally not directly affected by the problems they cause. Phytosanitary problems affect ecosystems and domestic producers who have no power to impose phytosanitary measures themselves. Therefore, phytosanitary measures are taken by governments in the interest of the country as a whole.

For the other types of standards (i.e. food safety, environmental and social) there is a high degree of interaction between corporate, NGO and regulatory standards.

This interaction means that any analysis of private sector standards needs to take into account the regulatory framework. In general, governments set minimum regulatory standards. Food retail companies that compete mainly on quality will want to position themselves with a quality level above the regulatory minimum. As retail companies govern the supply chain, they will try to impose the highest possible standard on their suppliers. Innovations will be picked up soon by other retailers that also compete on quality. For areas that are of public interest, such as food safety, governments may then be tempted to raise minimum regulatory standards, forcing those companies that compete mainly on price to also adopt higher standards.

Such a "race to the top" would depend on the willingness to pay of consumers in the higher quality segments of the market. This would in turn depend on the perceived differences in quality between the minimum regulatory level and the "high quality standard" (hence the role of labelling in the case of credence goods).

In the case of food safety (see example below), the proliferation of private standards would stop if the regulatory requirements were close to the maximum level of food safety that is technically possible to achieve, as the perceived quality differences would be minimal. Therefore, technical innovations create an enabling environment for the emergence of higher food safety standards. Because food safety research is often publicly funded, this is yet another level of interaction between private standards and government.

The interaction between technical regulations and private standards can be illustrated by the three examples below.

Food safety standards

Minimum food safety requirements have existed at the governmental level (marketing standards) for a very long time and they have been to a large extent harmonized through the Codex Alimentarius.

The last decades have seen an accelerating trend of concentration and globalization of the food retail sector. Large-scale retailers have become the most powerful actors in the food supply chain and the importance of retailer-own-brands has increased at the expense of other (supplier) brands. During the same period, food scares have given rise to more stringent liability legislation (first in the United Kingdom in 1990, then in the European Union in 1999). Because retailers are liable for their own brands, they have had a powerful incentive to develop private food safety standards to prevent food safety scandals. The certification schemes benchmarked by the Global Food Safety Initiative are predominant. They often cite Codex standards, notably the HACCP guidelines.

However, legislation in the European Union (General Food Law, 2002) is now catching up with the private standards. The new EU law also makes HACCP implementation compulsory for food packers and manufacturers (implementation foreseen in 2006). Also the US legislation is becoming more stringent as a result of the Bioterrorism Act of 2002, but mainly in the form of registration and trace back requirements and an increase in the level of import controls. Furthermore the FDA has initiated the Action Plan to Minimize Foodborne Illness Associated with Fresh Produce Consumption.

Organic agriculture standards

The first organic agriculture standards were developed by producer associations. With the expansion of the market, labelling became necessary to convey the information on the organic production methods over greater distances to the final consumers. With the proliferation of organic standards and labelling and the occurrence of false claims, the organic movement (producers, traders and consumers) requested governments to develop legislation. Proliferation of national standards in turn led to the request for the Codex Alimentarius Commission to develop international guidelines.

However, numerous public and private organic guarantee systems and regulations still pose challenges to international trade. The intergovernmental organizations FAO and UNCTAD, some national governments and the private organic farming sector (farmers, traders, certification bodies and NGOs) represented by IFOAM, work together in the International Task Force on International Harmonization and Equivalence in Organic Agriculture.

Labour standards

The standard SA 8000 developed by the NGO Social Accountability International is based on the core conventions of the International Labour Organization. The Belgian Government recognizes the SA 8000 standard as equivalent to the requirements of its social label and allows SA 8000 certified companies to use its label.

Process standards tend to be prescriptive instead of results based

Results-based standards state the results that have to be obtained, but let the implementing companies choose how to achieve these results. By contrast, prescriptive standards set precise requirements for how products should be produced. Such prescriptive requirements tend to pose more difficulties for producers in other production systems than those for which the standard was originally developed or with which the authors of the standard are familiar.

By default, product standards are more results based than process standards. Some prescriptive clauses in process standards are difficult to avoid (e.g. the prohibition of the use of synthetic pesticides in organic agriculture). However, process standards could be more results based than often is the case. For example, many food safety oriented standards aim to create hygienic production environments. Yet, instead of prescribing the desired result, they prescribe the means to achieve such results, to such details as the number and type of toilets that have to be available at a food processing facility.

Adherence to product standards can often be easily verified by looking at the product (e.g. for grading standards). Verification of adherence to process standards is however more difficult. That is why certification companies require extensive documentation in addition to the inspection of the production facility. Many standard developers already prescribe documentation requirements in the standards themselves. This makes it difficult for certification bodies to be creative in situations where documentation is problematic (e.g. due to high illiteracy rates). Overall, the need for documentation tends to make process standards more prescriptive.

Compliance with private standards is more relevant for exports to the European market than for exports to the US market

The difference between the buyer surveys commissioned by the World Bank is notable. In the report of the US buyer survey, Lamb et al. dwell on the challenges to comply with SPS measures implemented by the United States. In the report of the EU buyer survey, Willems et al. focus instead on private standards. What emerges from the literature is that the main constraint to exporting fresh produce to the United States is the governmental phytosanitary control system. The main constraint to exporting to the European Union is the proliferation of private standards.

In the United States, phytosanitary regulation is strict and strictly enforced, and therefore this remains the main entry barrier for many tropical fruits and vegetables from developing countries. An example is the obligatory heat treatment for mangoes and the frequent controls by US officials of heat treatment facilities in exporting countries. On the other hand, food safety measures are mostly in the form of voluntary codes. However, after an outbreak of food borne illnesses a particular origin may be banned until the problem is solved to the satisfaction of US officials.

US food retail companies require that suppliers comply with the law but unlike European retailers they have not developed standards themselves. The standards that have been adopted by US retailers have been originally developed by other stakeholders (for example organic standards were originally developed by producers) or in other countries (such as SQF).

In the European Union, the enforcement of phytosanitary regulations is reported to be more relaxed. Maximum residue level regulation has become strict, but the level of control is still reported to be low, albeit increasing. With entry possible at the regulatory level, suppliers face private sector demands. Partly to cover their liability risk and partly to protect their reputation, many retailers have developed requirements related to food safety that go beyond regulations or anticipate new regulations coming into force in 2006 under the new Food Law. European retailer associations have themselves developed standards for food safety and Good Agricultural Practices.

Arguably, the new EU food safety legislation, which is being phased in gradually, may change the situation. For those market operators that have not adopted the current private standards, minimum hygiene requirements for food packaging and processing will become more stringent. However, the actual effect will depend on the level of enforcement.

Private standards with the highest potential impact on market opportunities for developing countries are the Global Food Safety Initiative, EurepGAP and to a lesser extent organic agriculture.

The Global Food Safety Initiative

The Global Food Safety Initiative is important because it has a global scope, comprising both the US and EU markets. Retailer members of the GFSI told the OECD that they already require some food safety related certification from their developing country suppliers (either a GFSI benchmarked standard or EurepGAP). However, it should be realized that there are in effect 4 standards benchmarked against the GFSI: BRC, IFS, SQF1000/2000 and the Dutch HACCP code.

Although the standards are the same in essence, there are minor differences in specific requirements and in the competences required from auditors (see chapter 3 of Part 1). The differences in the requirements for the certification system prevent mutual acceptance of certificate equivalence. Suppliers may therefore still need multiple food safety certifications. However, according to an OECD survey, most retailers would prefer to have a single global food standard and certification system for food safety. This would decrease certification costs for the suppliers and allow retailers to switch suppliers and source across the globe.

The GFSI will have direct impacts on processors and packers of fruits and vegetables. Indirectly, it will affect primary producers through changes in sourced quantities or through changes in the sourcing patterns of the processors. The impact on market opportunities will be negative if the processors and packers cannot meet the standards. The impact may be positive if they can meet the standards and certification now enables them to convince buyers of the quality of their products (whereas they could not in the past).

EurepGAP

Whereas most retailers require a certification against a GFSI benchmarked food safety standard from their manufacturer suppliers, they increasingly require EurepGAP from the primary producers. At the time of writing, 27 food retailer groups in Europe supported EurepGAP. This affects directly small producers in developing countries. Having been developed in 2000 initially from a European perspective, and being highly prescriptive in nature, EurepGAP has a great potential to become a barrier to market entry. Only now an interpretation for small producers is being developed in Kenya. However, as with the GFSI, the impact of EurepGAP may be positive if it enables producers to convince buyers of the quality of their produce.

Organic certification and labelling

Developments in organic agriculture give conflicting indications as to how much it will affect market opportunities for developing country producers and exporters. The market for organic products is still expanding rapidly, whereas for most conventional food products the market grows very slowly. In certain markets for certain organic products, market share can be as high as 10 percent (e.g. for certain fruits and vegetable in the United Kingdom). However, globally, organic products constitute only 1 to 2 percent of the food market. This indicates that while organic standards presently have an impact on only a small proportion of farmers, the situation may change in the long run.

Organic labelling can have a positive impact on market opportunities for developing countries. Organic methods often require less capital (which is often scarce in developing countries) and more labour (which is often abundant in developing countries), giving developing countries a comparative advantage in organic agriculture. For producers in highly intensive production systems in developed countries it is more difficult to convert to organic methods, giving low-input producers in developing countries a comparative advantage

However, organic methods are often knowledge intensive, requiring high levels of literacy and good extension and education systems, which gives developing countries a disadvantage relative to developed countries. The proliferation of organic certification regulations, standards and labels increases certification costs and make switching markets more difficult. This problem is compounded for developing country producers who often have to pay more for auditor travel and for whom other transaction costs (translation of documents, etc.) may be higher too.

Overall, organic certification is likely to have a significant impact on market opportunities for developing countries: firstly, because the biggest and growing markets are the developed countries and there still exist supply constraints in some of these markets; and secondly, because organic products are sold at a premium.

Traceability requirements

Although not a standard per se, traceability requirements from the private sector may also have a substantial impact on market opportunities for developing countries. Both the United States and the European Union work towards regulatory implementation of one-step-back, one-step-forward traceability requirements. However, to enable swift recalls in case of a food safety failure, food retailers are increasingly implementing 100 percent traceability systems from the primary producer or even agricultural input suppliers to the final destination. New technological developments make this possible.

Although other certification programmes may affect fresh produce exports from developing countries, their current impacts can be considered as low.

Sustainable Agriculture Initiative Platform

It is difficult to assess the potential impact of the SAI Platform at first sight. The platform consists of major food manufacturers who form an important market for primary ingredient producers. However, the guidelines for potatoes and vegetables have been published only recently and the guidelines for fruits are still being developed.

The guidelines are developed with inputs from suppliers, i.e. the primary producers who have to implement them and go through a pilot testing process. Member companies do not seem to intend to adopt third-party certification programmes. Rather, they communicate the guidelines to their suppliers through business-to-business contacts as a part of their normal in-house quality assurance efforts.

All this together indicates that the implementation schedules are rather flexible and should not cause trade disruptions.

SA8000

This standard was developed for larger companies and does not involve a label on the product. The market share of the fruit and vegetable firms presently certified (Dole and Chiquita) is not directly affected by this certification, even though there may be indirect effects through reputation.

Rainforest Alliance

The Rainforest Alliance certification of bananas covers a large share of total Latin American production, but this is due to the certification of Chiquita plantations. Chiquita may have benefited from enhanced reputation, but their signature of the agreement with the ICFTU seems to have had more effect in this respect. The impact of the certification on the banana market may increase now that Chiquita is introducing Rainforest Alliance labelled bananas into Europe. Standards for citrus have been developed and a few firms have been certified, but this has not expanded further. No Rainforest Alliance standard exists for other fruits and vegetables.

Fair-trade

Except for bananas in the United Kingdom and Switzerland, fair-trade labelled fruits and fruit juices have an extremely low market share. Many fruit standards have just been developed and fair-trade labelled fruits and juices have just appeared or still have to appear in markets. Standards for vegetables have yet to be developed. Because of this low market share, fair-trade presently affects just a few fruit producer groups in each product category. For this reason, although fair-trade may be growing fast, organic agriculture will have a bigger impact for years to come.

ISO 9000 and ISO 14001

These standards have been mainly implemented by large companies and have become default certifications. Because of lax interpretation by many implementing companies and certification bodies, these certifications have got the name of being not more than a paper exercise. Thus, they have reportedly lost value in business-to-business relationships and consequently are less an advantage in marketing. Evidence that the absence of such certification could lead to exclusion from buyer contracts could not be found.

OHSAS

Because OHSAS is a new standard, no information on its implementation was found at all. As a management system that was modelled on ISO 9000, it could well have the same fate.

NutriClean

This label is only used in the west of the United States. Currently, its market share is very low.

2. Recommendations for future research

The overview presented in Part 2 of this report gives an idea of the directions of recent research by academic and other institutions. The impact of private standards on trade is a rapidly evolving area of analysis. Ongoing developments in global fresh produce supply chains, including the concentration in the food retail sector and the proliferation of voluntary and private sector standards, continue to give rise to new research questions.

Some topics have been more researched than others. There is, for example, a rapidly growing body of research taking a supply chain (or a value chain) approach. Case studies and analyses of specific national export industries or specific supply chains have already yielded many insights on the nature of barriers and the nature of the costs of compliance and benefits. It has become clear that the cost of compliance is farmer specific, as it depends on the initial production methods applied, the institutional and financial support received, the access to credit and information, the ecosystem and climate and the available infrastructure such as laboratories and telephones. More case studies are underway in FAO and various other organizations. Further case studies may still be useful for specific farmer groups or industries or specific standards not formerly studied. It should however be expected that they would yield few additional insights that would be generally valid.

The papers on methodology listed in section 2 of Part 2 indicate that economists have yet to solve many methodological problems before they can quantify the impacts of private standards in a meaningful way. The lack of large and reliable data sets poses another problem. The development of methodologies and data sets will have to go in tandem. Data collection will need to be guided by the developed methodologies, and methodology development needs to take data limitations into account.

Some attempts at modelling the impact of technical regulations (such as Maximum Residue Limits, MRLs) that are gradually phased in have been undertaken. However, most work has only been done at the theoretical level so far.

For a better understanding of the impacts of private standards on market opportunities for developing countries, further research is necessary in the following areas:

i. Quantities of products certified against private standards
ii. Distribution of compliance costs and value added along the supply chain through price transmission.
iii. Consumer willingness to pay higher prices for certified and labelled foods
iv. The impact of private standards on the competitive position of fresh produce industries in developing countries.
v. Policy options and the impact of different policy choices as regards private standards
vi. The effectiveness of different approaches to technical assistance
vii. Implications of the GATS agreement for standard setting and conformity assessment services

Each of these areas is shortly discussed below.

i. Market share

An understanding of the percentage of trade in fresh produce that is governed by private standards would be a logical basis for the analysis of the impact of these standards, yet this information is not readily available. Only quantitative studies on the organic and fair-trade markets are available. This is not very surprising, as most of the other standards and certification programmes do not provide for product labelling. Due to the absence of label on the products, the market that is targeted consists of corporate customers (in particular retailers) rather than final consumers. The volume and value of certified produce purchased by retailers are not published. Furthermore, this information may be considered to be commercially sensitive.

Retailers may publicly claim that they require a certain certification while they continue to buy uncertified produce. They may not demand certificates from trusted suppliers, but it is not in their interest to disclose this information for fear that other suppliers might refuse to invest in certification.

Therefore, reliable figures on the market share of these standards are not readily available. Obtaining reliable estimates would greatly improve the quality of further analysis. For example, many authors claim that some private standards are becoming de facto compulsory for market entry. Such claims could be validated by data on the share of the market actually governed by these standards.

ii.   Distribution of compliance costs and/or consumer price premiums along the supply chain through price transmission

Supply chain studies suggest that cost distribution is minimal for food safety standards that are imposed on producers: additional production costs are not compensated by higher farm-gate prices.

The study by Kilian et al. (2005) suggests that the change in revenues resulting from conversion to organic or fair-trade standards varies along the marketing chain. At a given point in the chain the price premium may not or only just cover the compliance costs, while at other points in the chain they more than offset extra costs.

Again, the lack of reliable data hampers such analysis. FOB and wholesale prices are often available, but because retailers increasingly buy directly from importers or exporters, wholesale prices cover a diminishing share of the market. Price data of direct transactions are commercially sensitive and actual prices paid may differ from officially published prices due to pay-back clauses in contracts or promotional actions.

To analyse margins, price data have to be complemented by data on costs. Studies on the cost of compliance show that the costs generated by standard and certification are highly variable, depending on origin and firm specific situations. The costs incurred by retailers (extra shelf space, etc.) are seldom available. The analysis of cost/premium transmission is further complicated by rapid supply and price fluctuations, due to the seasonal and perishable nature of fresh fruits and vegetables.

iii.   Consumer willingness to pay

An increasing number of studies use experimental auctions to determine the consumers' willingness-to-pay for certain credence goods, or the discount required in the case of negatively perceived aspects (e.g. GM food). A methodological discussion and two examples are included under section 8 of part 2. Despite some methodological issues, these experimental auctions are believed to give more reliable results than surveys based on interviews. They could therefore be an important tool for determining the market potential of labelled certified products marketed at higher prices. However, no auction of organic versus conventional fresh fruit and vegetables was found in the literature surveyed by the author.

iv.   Analysis of the impact of private standards on the competitive position of fresh produce industries in developing countries

The above overview includes several studies on the effects of standards on a particular industry in a specific country. They usually focus on the (lack of) investments made resulting in the success or failure to enter markets governed by standards and on related structural changes in the sector.

Some larger research projects have done several of such industry studies and then tried to identify some general patterns. However, no comparative studies on the same industry across many countries have been undertaken so far (e.g. a cross-country study of the investments in the pineapple sector in reaction to EurepGAP).

It would also be interesting to analyse the market shares of different origins before and after the implementation of a standard. This would give an indication of which countries have managed to take advantage of new standards. However, it would tell little about the causal relationship and the causes for this success. Since most private standards (EurepGAP, BRC, SQF etc.) have only been developed recently and their adoption is still ongoing, it may be too early for such an analysis.

v.   Policy options and the impact of different policy choices to deal with the rise of private standards

Many papers provide policy recommendations and other papers even focus on policy options in reaction to emerging private standards. However, no study was found on the effect of different policy choices on export opportunities for farmers. For the organic farming sector, enough data may be available to make such an analysis.

Exports from countries with a similar potential but different policy responses could be compared. Potential research questions could be: do developing countries that have developed a national organic legislation have a better position in organic export markets than countries that have not? What has been the impact of setting up national certification bodies? What has been the effect of training programmes on organic agriculture for extension agents?

vi.   The effectiveness of different approaches to technical assistance

Technical assistance can be provided at governmental level, at the level of auditing, certification and accreditation services, at the level of industry association and at the farm level. Technical assistance can also have various aims, such as increasing the negotiation capacity of countries or farmer associations, increasing standard implementation capacity or linking farmers to export niche markets for labelled products. An analysis of the effectiveness of technical assistance should clearly take into account these different levels and objectives.

The author found very few analytical studies that evaluate different forms of technical assistance in the horticultural sector. As many private standards have been developed only recently, related technical assistance is recent. Most experience has been gained so far in assisting farmers in meeting the requirements of import regulations and of certification programmes for organic and fair-trade markets (especially for coffee, banana and cotton). The number of projects that give assistance to farmers in the adoption of private standards is growing.

vii.   Implications of the GATS agreement for standard setting and conformity assessment services

There are various studies on the implications of the TBT and SPS agreements for standard setting and conformity assessment in the agricultural sector. Some specific issues may still need further analysis, such as Article 4.1 of the TBT Agreement on the reasonable measures that members are expected to take to ensure that non-governmental standardizing bodies within their territories accept and comply with the Code of Good Practice for the Preparation, Adoption and Application of Standards.

In section 1.1 of Part 1 of this report, it was mentioned that some authors argue that standardization and conformity assessment are a service industry in itself and would fall thus under the GATS agreement. The studies related to standards and GATS that were found deal with standards in service industries such as education, tourism or air transport. The author of this report could not find any study that provides insight into the implication of GATS for standard setting, certification and accreditation bodies per se. It would be useful to contract international trade law experts to study the implications of GATS for the conformity assessment industry. The findings would be relevant to standard setting and conformity assessment in agriculture.